Aircraft Systems

Other books in the PRACTICAL FLYING SERIES

Aircraft Systems

Second Edition

David A. Lombardo

McGraw-Hill

New York   San Francisco   Washington, D.C.   Auckland   Bogotá
Caracas   Lisbon   London   Madrid   Mexico City   Milan
Montreal   New Delhi   San Juan   Singapore
Sydney   Tokyo   Toronto

Library of Congress Cataloging-in-Publication Data

Lombardo, David A.
    Aircraft systems / David A. Lombardo.—2nd ed.
      p. cm.
    Includes index.
    ISBN 0-07-038605-6 (alk. paper)
    1. Airplanes   2. Private flying   I. Title.
  TL670.L66   1998
 629.134—dc21
                                          98-31966
                                              CIP

McGraw-Hill

*A Division of The **McGraw·Hill** Companies*

  2 3 4 5 6 7 8 9 0 DOC/DOC 9 0 3 2 1 0 9

ISBN 0-07-038606-4 (HC)
ISBN 0-07-038605-6 (PBK)

The sponsoring editor for this book was Shelley Ingram Carr, the editing supervisor was Sally Glover, and the production supervisor was Sherri Souffrance. It was set in Times per the PFS design by Michele Zito and Kim Sheran of McGraw-Hill's Professional Group Composition Unit, in Hightstown, NJ.

Printed and bound by R. R. Donnelley & Sons Company.

To Dad
and
The Nerka

Acknowledgments

This second edition is a complete revision and restructuring, so there are many people to whom I am indebted for their assistance. I must warmly thank Rich Mileham, airworthiness inspector, Great Lakes Region of the FAA. Rich, while an inspector at the Chicago-area DuPage FSDO, thought it didn't make much sense that pilots had accident prevention counselors but mechanics didn't. Tilling sacred ground, Rich was generous enough to appoint me to the first group of FAA airworthiness accident prevention counselors. Over the following years as we traveled throughout the area conducting safety seminars for maintenance personnel, I came to have an enormous respect for this quiet, innovative Fed with a sense of humor. When, in the course of having this revised edition written, a couple of reviewers fell through, Rich filled in with lightninglike turnaround time and some excellent input. Rich Mileham is the quintessential public servant and a good friend.

I would also like to thank my very good friend Rudy Frasca, president, Frasca International, Inc. for his long-time support of my writing efforts and for making his wide array of aviation facilities available in my quest to add new photographs to this book. I would especially like to thank Tom Frasca, manager, Frasca Air Services for taking so much of his time to help me take photographs for this edition and Dale Rapp of the Frasca Air Museum for his assistance and historical knowledge.

For their assistance in providing fresh illustrations and photographs, I would also like to thank David Henry, chief pilot/engineer, Aerospace Systems & Technologies, Inc.; Jennifer Whitlow, manager marketing communication, Cessna Aircraft Company; Bruce Rivard, vice president, Lake Aircraft, Inc.; and Kimberly V. Wheeler, manager of marketing, The New Piper Aircraft Corporation. When other, giant manufacturers either weren't interested or couldn't respond with turnaround times of less than several months because of restrictive legal departments, AS/T, Cessna, Lake, and Piper were shining lights.

ACKNOWLEDGMENTS

It is a telling tale that so many manufacturers are so bogged down in fear of litigation that they cease to be dynamic and openly supportive of pilot education. With only a few exceptions, it is the smaller, entrepreneurial companies that took a strong, active interest in this book. I urge the reader, whenever possible, to support the above and following companies, all of whom displayed a strong desire to help educate general-aviation pilots. They are willing to help you; please help their businesses grow and prosper.

For reviewing the content of individual chapters I am indebted to the following professionals for taking their time in support of this project: Jonathan Bent, VP marketing for De Vore Aviation Corporation of America; Vincent Marotti, business development manager for AIL Systems Inc.—Dorne & Margolin Antenna Technologies; Debbie Abrahams, inside sales manager and marketing and old friend Joyce Ervin, applications specialist, both of Micro-Surface Finishing Products, Inc.; Frank T. DeAngelo, president of ARC Industries, Inc.; Lou Bisonni, customer service department, Kollsman, Inc.; Shawn Ewing, technical service specialist, Phillips 66 Company; John Thurston, manager of technology and operations, Air BP; Tom S. Johnson, aviation manager—general-aviation sales, Texaco, Inc.; Dick Vincent, AlliedSignal Turbocharging Systems; Gene Martin, general manager, Martin Induction Systems; Michael R. Disbrow, VP marketing and customer support, and Rick Bowerman, chief applications engineer, for Hartzell Propeller, Inc.; Harold "Hal" Haskins, Jr. general manager, Harold Haskins, Inc.; Terry Holburn, VP marketing of Keith Products, Inc.; Pike Kelly, program manager, Dukes, Inc.; Mark Gaulke, technical sales, Rapco, Inc.; Bryn E. Young, senior product support representative, Ice Protection Systems Division of BFGoodrich Aerospace; David Downer, sales manager, Hydraulic Pump/Motor Division of Parker Hannifin; Bruce Moncreiff, manager—hydraulic fluids, Castrol Industrial, Inc.; David Smith, chief engineer, Allenair Corp.

I must continue to thank Vicki Cohen, who for many years put up with my quirks and idiosyncrasies. Rejoicing with me in the good times and helping me through the tough ones, her support helped me make the transition from employed slave to self-employed entrepreneurial slave. "Mom" to our puppies, she is still my closest friend and I will be indebted to her forever.

And finally, I thank those to whom this book is dedicated: Dad and The Nerka. Sir Roscoe of Inverness, affectionately called The Nerka, passed away during the writing of this edition. My best friend and faithful golden retriever of ten years, his passing leaves a huge void in my life and the life of his widow, Molly Inverness. Not a day goes by that we don't think of, and miss, him. The love and emotional support he selflessly provided all those years can only be appreciated by someone who knew him.

Dad has passed away since the writing of the original book but lived to see it published. I never saw him prouder than the day he saw his son's name in print somewhere other than the local newspaper's police blotter where so many expected to see it eventually. Through good times and bad, he always encouraged and supported my every endeavor. No child could have asked for more love, no adult for more approval. Through 10 years of college and countless moves, his timid question was only, "where is this all going?" And finally my answer was, "here."

Contents

CONTENTS

CONTENTS

Introduction to the Revised Edition

During the 10 years since this book was published, I've heard from a lot of people. *Aircraft Systems: Understanding Your Airplane* had become an aircraft systems text for many flight students. The original book was a collection of magazine articles I had written, with a few chapters written for the book. The comment I heard most often was that it was difficult to use in a classroom setting because of a lack of subject organization. This edition addresses the concern.

The original 31 chapters have been reduced to 15 subject-specific chapters. While this has resulted in relatively large chapters dealing with the airframe, the powerplant, the instruments, and the electrical system, this will make using the book as a text much simpler.

The material has been updated and in many cases completely rewritten for increased clarity. There have been 16 photographs and 9 illustrations added to help visual concepts, and the table on preventive maintenance has been updated. The new FAA-approved list almost doubles the things you can do to your aircraft to save a little on your maintenance bills.

Finally, I responded to comments by readers regarding areas that should be covered in the book. As a result, the new edition includes information on the following subjects, which were not covered in the first edition:

- Electromechanical actuators
- Powerplant theory of operation
- The development and chemistry of gasoline
- Avgas octane ratings and performance numbers

Introduction to the Revised Edition

- Engine oil analysis
- New SAE oil viscosity rating system
- Rust and corrosion
- Landing gear systems
- Anatomy of a tire
- Weeping wing ice removal systems

The improvements in the material and the coverage of this edition are directly related to comments I have received over the years from various readers. I highly encourage readers and instructors to send their ideas and comments to me for consideration in the next edition. By understanding what you need, I can continue to focus this book appropriately. Thank you for your interest and support.

David A. Lombardo
June, 1998
Shorewood, IL

Introduction to the First Edition

Since June 17, 1969, I have been continuously involved, either directly or indirectly, in aviation/aerospace education. On that day, I received my first dual flight in a Cessna 150, and although both the airplane and the airport are gone, the memory is as fresh as if it happened yesterday. Flying is something that gets in a person's blood; even an extended absence cannot prevent a person from saying, "I'm a pilot." But the love of flying should never be confused with the technique of flying.

Thousands of pilots have safe flights every day, but that doesn't mean they fly safely everyday. Many pilots assume that because they have both a large, and an equal, number of takeoffs and landings, they are safe pilots. Unfortunately, that is not the case, and NTSB accident statistics prove it. Safety is a relative concept; it is directly proportional to skill, knowledge, and judgment.

This book stresses knowledge of general-aviation aircraft systems. It isn't as if systems have been totally avoided by authors; there are some excellent texts for use in aircraft maintenance schools. Many universities and other professional pilot programs have been using these texts, but unfortunately there are several serious drawbacks. Such texts always go into far greater detail than is warranted, or even desired, for pilot education. The material often requires specialized knowledge of physics, mathematics, and chemistry not covered in the book. The necessary material is seldom available in a single text—typically requiring two or three. And in virtually every case, the price of each text is very high. I decided that a reasonably priced, pilot-oriented book was necessary.

Aircraft Systems: Understanding Your Airplane requires no special knowledge from the reader. When necessary, terms are defined and concepts are clarified. The book will be as useful to the student pilot as it is to the experienced one. Since the chapters are divided by aircraft systems, it may be read from cover to cover or used as a reference book.

Introduction to the First Edition

For quite a few years I have been a very active aviation safety public speaker and volunteer FAA Accident Prevention Counselor. Recently I have been faced with the awful truth that despite the tremendous effort made by the FAA and other aviation safety-oriented groups, it's not possible to reach all the pilots through safety seminars. It is my deepest hope that this book, and others like it, will be read by every general-aviation pilot. Knowledge is the greatest tool a pilot can ever possess.

David A. Lombardo
November, 1987
Champaign, IL

1
The Airframe

SOMEONE ONCE SAID, "DON'T MEET TROUBLE HALFWAY; IT IS QUITE capable of making the entire journey." So it can be said for maintaining an aircraft. Owning and operating an aircraft is in no stretch of the imagination inexpensive. The time and money you spend on preventive and routine maintenance, as well as conducting a thorough preflight inspection before every flight, can be substantial, but the long-term payoff is a safer airplane with lower overall maintenance costs.

THE PREFLIGHT INSPECTION

It would be difficult to overemphasize the importance of a thorough preflight prior to every flight, and that definitely includes an exterior airframe inspection. I have observed pilots on countless occasions using a checklist to do the cockpit checks but not using it for the exterior portion of the preflight. When I've asked them why, I've never had a single pilot give me a good reason for eliminating the use of the checklist on the walkaround preflight.

As tedious as it may be, using a checklist will guarantee that every item is inspected. But realistically, few civilian pilots will actually pull out the checklist and conduct an exterior preflight inspection. Therefore, if you are going to ignore the safest practice, it is essential to at least conduct every walkaround inspection in exactly the same manner.

CONDUCTING THE EXTERIOR PREFLIGHT

When doing an exterior airframe preflight inspection, establish a pattern. Start at the same place every time and move in the same direction around the aircraft. In lieu of using a structured checklist, a pattern will help reduce the possibility of missing something. Typically, if you get a "funny feeling" something is wrong when you use a pattern, you're subconscious is telling you that you have missed something. Trust your feelings; go back and take another look.

First, as you approach the aircraft, look at it from a distance. Does anything look unusual? I know of two instances where pilots have attempted to start an aircraft engine when there was no prop. How can you miss a prop on a preflight inspection? And I once watched a student pilot conduct an exterior preflight on an Aeronca Champ that had one wing's angle of incidence (the angle between the chord of the wing and the centerline of the aircraft) significantly greater than the other. Up close to the aircraft, the student couldn't see both wings at the same time, so the difference was not readily observable. The airplane would have had a potentially uncontrollable rolling tendency in flight. The problem was a fuel truck had accidentally pushed against one wing, causing its leading edge to shift upward. The lesson learned is to look at the aircraft from a distance and see if everything appears proper.

Three Areas of Consideration

When you arrive at the aircraft there are three areas of consideration during an exterior preflight: on the ground, on the surface, and under the surface. On the ground relates to things you see directly below or around the aircraft. On the surface is about covering normal preflight items—things that are visible even if you have to open engine access doors to find them. Finally, under the surface refers to those things that may affect the health of your aircraft that sometimes occur out of sight or are not so easily determined—such as rust and corrosion.

On the Ground

Before beginning the actual walkaround inspection, observe what's on the ground underneath and around the aircraft. Are there loose parts? If so, did they come from the airplane or are they just foreign objects? Always keep the ramp clean, as foreign object damage (FOD) can easily result to your aircraft, other aircraft on the ramp, or airport facilities when a turning prop hurls them.

Is there a pool of liquid underneath the aircraft? Is it fuel, oil, or hydraulic fluid? Look directly above the pool and see if there are telltale traces on the underside of the aircraft. Overfueling an aircraft can cause spillage. Also, fuel will expand on a hot day and run out of the fuel tank vents if the tank has been filled to the top. It is important to determine if the fuel on the ground represents a leak. Except for spillage while trying to replenish oil or hydraulic fluid, there is no reason for either to be under an aircraft. Don't take any leaks for granted.

On the Surface

When reviewing "on-the-surface" issues, there are two primary considerations: aircraft security and attrition. Aircraft security relates to the status of the aircraft's physical condition. The preflight checklist essentially assures that everything is where it belongs and is in working condition. But there should be more to the preflight than just checklist items. For instance, is the aircraft clean?

Aside from aesthetics, there are practical reasons for a clean aircraft. You can see telltale traces of fluid leaks, popped rivets, and other problems more readily on a clean aircraft. Spotting popped rivets is fairly simple, as black oxide seeps out from under the rivet as illustrated in Figure 1-1. Simply sight along a line of rivets and look for small, black stains. If you find one, press the surface next to the rivet to confirm it's loose. Popped rivets are indicative of potentially serious damage such as an overstressed wing spar, so always notify a mechanic.

Fluid leaks show up better on clean aircraft and often can be traced along the fuselage to their source. Another benefit of a washed and waxed airplane is that it will be aerodynamically cleaner, resulting in an increase in your cruise speed of a couple of knots.

Speed fairings on fixed-gear aircraft are another item that should be checked. They're the things that your mechanic curses when it's time to change a tire. Folks up north call them snow catchers and usually remove them permanently, the first snowfall after they buy the airplane. Not only do they increase cruise airspeed, but many cruise performance charts such as those in the Cessna 152 pilot's operating handbook (POH) are calculated with the fairings on. Read the fine print closely, and it will tell you if the computations reflect them.

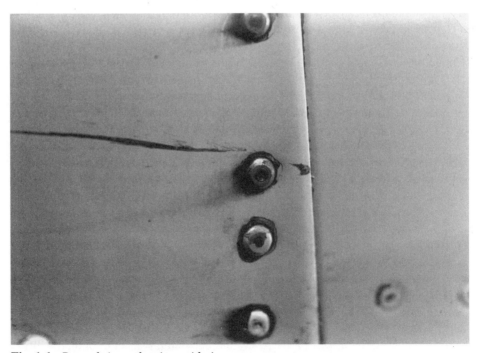

Fig. 1-1. *Popped rivets showing oxidation.* (Photo by author, courtesy of Frasca Air Services)

Chapter One

Attrition, the second primary consideration during preflight, relates to the general health of the aircraft. There are five aging elements that impact long-term health. They are: weather, friction, overload, heat, and vibration.

Weather constantly erodes the airplane. Changes in temperature and humidity have negative effects on both the exterior and interior of the aircraft as well as aircraft instruments. While rain, snow, ice, and especially hail can occasionally have a dramatic, short-term effect on an aircraft exterior, it is the subtle, long-term weather effects that generally take the greatest toll. Wind, sun, and dust constantly chisel away at aircraft paint and windows.

Friction, the wear between moving parts, is another insidious problem. Checking control surfaces to assure that they move freely and in the proper direction doesn't say much about how worn the hinges have become. Aileron, elevator, and rudder hinges should be scrutinized not only for freedom and proper direction of travel but also for excessive play. Hinges should also be visually inspected to assure that there is no damage or visible wear. And yes, I did say check for proper direction of travel. I once had a student preflight an airplane just out of maintenance, and he came to me and questioned the direction of travel of the ailerons versus the yoke. After I was done chiding him for not knowing, he told me that he was pretty sure that was right but the airplane wasn't responding that way. The cabling, which had been worn, was replaced and incorrectly rigged during maintenance.

There are also less obvious areas of friction that are nonetheless important. Check places where two things come together even if they are not intended to be a point of relative movement. I have seen cooling baffles that have rubbed their way through an engine cowling. It is far less expensive to resecure or replace an engine baffle without having to repair or replace the cowling at the same time.

Some overloads are fairly easy to detect, as they show up as cracks. If they are in a structural part of the aircraft, you must consult a mechanic. On the other hand, cracks in nonstructural parts such as fairing and wheel pants do not necessarily make an aircraft unairworthy. Look at the crack and determine if the relative wind in flight will cause it to open. Breaking off a piece of fairing or having a severe airflow disruption could not only be a real shock to the pilot, but also could result in aerodynamic problems. A neat, single layer of duct tape may be used to secure small cracks in nonstructural areas, but a good rule of thumb is to have it looked at by a mechanic before flight.

Other overloads are difficult, if not impossible, to detect after the fact. Severe overloading may result in deformation and structural failure. An overload can be induced in a number of ways. One sure way is to do a steep banked turn, sharp nose over, or other high-G-producing maneuver when the aircraft is loaded beyond allowable limits. The airframe may also experience overload conditions, even though significantly below maximum gross weight, while encountering turbulence if the airspeed is above maneuvering speed or during a close encounter with a thunderstorm.

Student pilots will tell you that "Any landing you walk away from is a good landing," but that's a matter of perspective based on who's paying the aircraft's maintenance bills. Probably the most common way of exceeding the maximum load limit of an aircraft is hard landings. Most susceptible are those aircraft with low wings whose landing gear are attached to the wing spar. The excessive stress causes the wing spar to bend upward, which stretches the skin under the wing. Wrinkles in the aluminum skin on the bottom of the wing are a telltale sign. And once stretched, the skin never completely returns to normal, so you can slightly flex the skin by pressing on it, much like pressing on the side of a gas or oil can; hence the name "oil canning."

Student pilots notwithstanding, some "give" in the surface is to be expected, but considerable flex may indicate severe damage to the wing and spar. Another sign of the same type of damage would be popped rivets on either the top or bottom of the wing. Trainer aircraft experience a higher than average incidence of hard landings, so you should look for popped rivets and oil canning first when considering renting or purchasing a low-wing trainer aircraft. The potential damage resulting from hard landings is one of the reasons why I would personally never buy a used aircraft that had been used extensively for pilot training.

Heat provides both direct and indirect concerns for the aircraft operator. Directly, carbon monoxide can leak into a cabin because of a faulty weld in the aircraft exhaust system. Exhaust welds should be periodically checked, and all aircraft should have an inexpensive carbon monoxide detector inside the cabin. Not only should the pilot frequently check the detector because the gas is colorless, odorless, and tasteless, but the detector should be replaced frequently. Carbon monoxide detectors do not have a long life span, so they should be replaced several times a year.

Indirect heat problems also tend to relate to engine operation such as inadequate engine cooling. Preflight and in-flight symptoms indicating potentially severe engine problems are high oil and cylinder head temps, the odor of burned oil or hot rubber during engine operation or shortly after shutdown, and auto-ignition of an engine after shutdown. Auto-ignition refers to a situation in which the pilot attempts to shut down the engine and it continues to keep running by itself.

One sure sign of an indirect heat problem is watching cowling paint over the engine blister during start. This is a sure sign of an induction fire. The average pilot will panic and stop trying to start the engine, which will significantly worsen the problem. Shutting down the engine during an induction fire allows the fire to burn inside the cowling, resulting in potentially major damage. What has happened is a fire has occurred in the engine's fuel/air induction system. Continued cranking of the engine will result in the fire being drawn back into the cylinders where it belongs. Make no mistake; once the fire is out and you do disengage the starter, the engine has a serious problem and needs to be referred immediately to a mechanic.

The last of the five aging elements is vibration. There are many normal vibrations in an aircraft which, over time, contribute to the aircraft's deterioration. Normal flight-related vibrations will loosen unsecured nuts and bolts over time, stress components, and contribute to gyro bearing wear to name a few. There really isn't anything you can do to prevent normal aircraft vibration, but unusual vibrations signal danger.

There are several typical causes of abnormal vibrations—for instance, ice disrupting the airflow over the airframe, a control surface, or antenna. A loose control surface will cause a noticeable vibration in the corresponding pilot control. Other causes include flying the aircraft at an airspeed in excess of the designed normal operating speed and vibrations stemming from engine or propeller-related malfunctions.

It is unlikely that a vibration-causing problem can be cured in flight. The best a pilot can hope for, in most cases, is to attempt to reduce the vibration by reducing power and/or airspeed. It is also important to avoid any increase in the load factor, so make very gentle turns, climbs, and descents. Take it as easy as possible and land the aircraft as soon as practical.

Under the Surface

The final consideration during the exterior preflight is what's going on under the surface, where you can't easily detect problems. Rust and corrosion are facts of life; all aircraft have it somewhere at one time or another. Rust is oxidation that occurs when ferrous metals such as iron or steel react with oxygen. The symptom is a reddish discoloration that, in its early stages of development, can often be totally removed simply by rubbing with a clean rag then applying a coat of wax to protect the area. If it is a bit more persistent, it may be necessary to rub the area with a synthetic fiber or stainless steel brush; never use a steel wire brush, as it will cause more problems in the long run. Once the rust is removed, it is necessary to paint it with a zinc chromate primer, then repaint to match the surrounding area. A reddish-brown crustiness indicates a more advanced development of rust. When you remove the surface rust, pitting will be evident in the metal. This situation requires a mechanic's evaluation.

A major problem associated with rust is that it often occurs in hidden areas where, left unnoticed, it easily develops to an advanced stage. Typical locations include in the belly of the aircraft, inside control surfaces, inside the wings and empennage, and inside steel-tube members of float planes. Aircraft that operate in wet environments regularly, or operate near salt water, should be routinely and frequently inspected by a qualified mechanic. This is particularly true for aircraft that have been converted into seaplanes or amphibians by removing the wheels and replacing them with floats. Those aircraft should be checked frequently because they didn't get the factory corrosion proofing always put on newly manufactured aircraft built for amphibious use, such as the Lake Renegade in Figure 1-2.

Corrosion is another form of oxidation that occurs when oxygen reacts with metal. For instance, it is aluminum oxide that causes shiny aluminum alloy surfaces to dull. Corrosion that occurs on a surface under the paint will become noticeable as surface flaking, pitting, blistering, or bubbles. In its early stages, corrosion may be removed with a gentle cleaning. In the advanced stages of corrosion, after pitting in the metal has occurred, corrective measures require the attention of a mechanic.

Corrosion can also be the result of an electrochemical process and is easily recognized as a grayish-white powder such as that which forms on the terminals of a battery. It involves dissimilar, nonferrous metals such as aluminum, copper or magnesium, where they directly contact one another or are indirectly connected through an electrolyte. Certain chemicals, such as battery acid or fumes, exhaust gases, and even acid rain will interact with dissimilar metals placed side by side. Other chemicals such as insecticides

Fig. 1-2. *The Renegade amphibian.* (Compliments of Lake Aircraft, Inc.)

and fertilizers are particularly troublesome to crop duster aircraft. Moisture, and especially salt water, trapped on a metal surface by dirt, mud, damp floor carpeting, or insulation, or the result of a plugged drain hole will also result in corrosion. Even hangared airplanes are susceptible, as bird droppings will cause the same problem.

Fundamentally, an electrical circuit is synthesized and the metal that offers the least resistance to corrosion becomes the anode resulting in its corroding. In general, corrosion is likely in areas where the surface is unprotected. Always touch up a surface scratch that has removed the paint down to bare metal. Another area conducive to the formation of corrosion is anywhere you can expect metal fatigue for such reasons as flexing, rubbing, and compression.

Areas particularly requiring frequent inspection include engine exhaust areas, landing gear, wheel well areas, the external cooling vanes on engine cylinders, surface skin seams, piano hinges on control surfaces and access doors, and battery compartments and vents. To neutralize a battery box in which there has been an acid spill, dust the box with baking soda, flush well with water, then thoroughly clean the box. When the box is completely dry, you can refinish it with zinc chromate primer. Make sure to coat the battery terminals with grease to retard future corrosion.

There are several things that the owner can do to minimize the potential for corrosion. Storing the aircraft in a hangar will go a long way toward protecting it. The airframe should also be washed and waxed regularly and thoroughly dried. An aircraft needs to be flown periodically to heat up the engine and air out the airframe. All electrical equipment should be periodically used to evaporate moisture. Fix any exhaust leak immediately, both for your own safety as well as to reduce the potential for corrosion. Periodically check the battery and fuel vent lines to be sure they're open and have a free flow of fresh air. Use a fluid, thin film coating spray penetrant on screws, rivets, and joints where practical to get rid of moisture, stop electrolysis, and inhibit corrosion.

AIRCRAFT LIGHTING

In the earliest days of flying, there was no need for any type of aircraft lighting. As aircraft became more practical to use, intrepid aviators began penetrating the night sky to go further and get more work done. But it wasn't until the 1920s that there was enough nighttime air traffic to begin to cause concern about seeing other aircraft. It was then that aircraft engineers turned once again to the sea to carry over some traditions to the sky.

Aeronautical position lighting took on the characteristics of its nautical cousin. The port wingtip was marked with a red lamp, the starboard with a green lamp, and the tail with white. They were clear, six-candlepower bulbs dipped in red or green lacquer as appropriate and, because in those days aircraft had no electrical systems, a 6-volt auto battery was carried onboard to power the lights.

Position Lights

Today, the descendants of those red, green, and white position lights are required by current regulations whenever operating during the period of sunset to sunrise. The requirement is somewhat different when operating in Alaska. There, position lights must be illuminated during the period when a prominent unlighted object cannot be seen from three statute miles or the sun is more than 6 degrees below the horizon. This same regulation also covers taxiing or parking aircraft and requires that the area be illuminated well, the area be marked by obstruction lights, or the aircraft have lighted position lights.

According to FAR Part 23.1385, which deals with position light system installation, the forward red and green lights should be spaced as far apart as practicable, typically on each wingtip. The lights face forward, with the red light on the left side and the green light on the right from the pilot's point of reference. The rear-position white light should be mounted as far aft as practicable, usually on the tail, though some aircraft have aft-facing white wingtip lights instead. The purpose of position lights is not simply to be seen at night. They also help the pilots of other aircraft determine your direction of travel, so position light field-of-coverage is carefully specified in the regulations.

The visible vertical pattern must be 180 degrees centered on the horizontal centerline, as illustrated in Figure 1-3. The horizontal pattern for the red and green lights is 110 degrees of coverage each, from the centerline, and 140 degrees for the white light. The phrase used by many pilots to help remember the proper orientation is "red, right, return," which means if you see an airplane with a red light showing to your right side, then it is traveling toward (returning to) you.

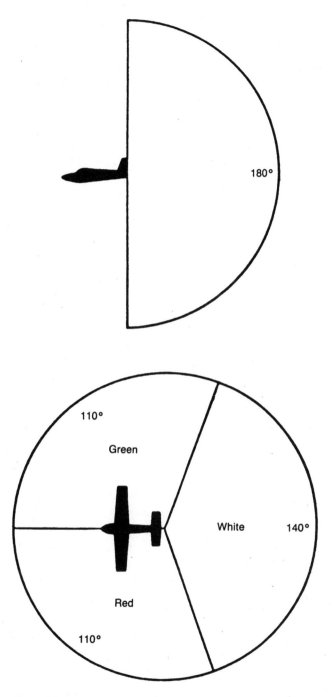

Fig. 1-3. *Position light pattern requirements.*

Anticollision Lights

The post-WWII era saw an economic boom in the United States, a large number of military-trained pilots, and many ex-military transport aircraft available for conversion to civilian use. By the 1950s these factors combined to produce a proliferation of airlines; the skies were more crowded than ever before in history, and for the first time airplanes were routinely carrying business travelers. The demand for greater aircraft conspicuity paralleled the demand for night airline operations, so anticollision lights were developed.

The earliest anticollision lights were white rotating beacons. It wasn't too long before it became obvious that a flashing white light resulted in a very significant, bright backscatter in the cockpit which could be very disorienting to the crew. As a result the white light was covered with a red filter to reduce the effect. Eventually the strobe light was developed and the requirement for an anticollision light broadened to include either a rotating beacon or a strobe.

The requirement for an anticollision light today is defined by four aircraft certification date categories. If the aircraft was certificated prior to August 11, 1971, it must have at least a red rotating beacon. If it was certificated on or after August 11, 1971, the regulations call for a white incandescent light, which is more luminous than the traditional red light. For aircraft certificated after July 18, 1977, the requirements regarding power and coverage for anticollision lights are more stringent yet.

While aircraft are grandfathered in, it is always acceptable to upgrade the older anticollision light systems for a more modern one. The simple fact is older collision light requirements are inadequate for today's dense air traffic environment. All aircraft can now be outfitted with simple, cost-effective strobes, which in most cases can utilize existing wiring and lighting mounts already on the airframe.

The current specifications for both rotating beacon and strobe anticollision light systems as defined by FAR 23.1401 may include one or more lights of either aviation red or aviation white. The intensity requirements are specified in detail by the regulation to assure adequate visibility. The configuration must be such that it "illuminates the vital areas around the airplane," which functionally means the light system must project light 360 degrees around the aircraft's vertical axis, as shown in Figure 1-4. It also must project light 75 degrees above and below the horizontal plane of the aircraft. Additionally, the system must produce a minimum of 400 candlepower in a forward direction, though the effective light-intensity requirement diminishes as the angle increases from the centerline (the regulation provides a table detailing the tolerances), and the lights must have a flash rate between 40 and 100 cycles per minute. While strobe lights are very effective, there are some drawbacks of which pilots should be aware.

Some people are highly susceptible to nausea, disorientation, and even the onset of epileptic episodes as a result of strobe lights. The regulations allow for the deactivation of strobe systems if the pilot in command determines that operating conditions warrant it. It is highly recommended that a strobe system be turned off when flying in clouds or fog, which causes the light to scatter around the aircraft and into the cockpit. It is also acceptable to turn them off during ground operations because the intensity of a strobe is

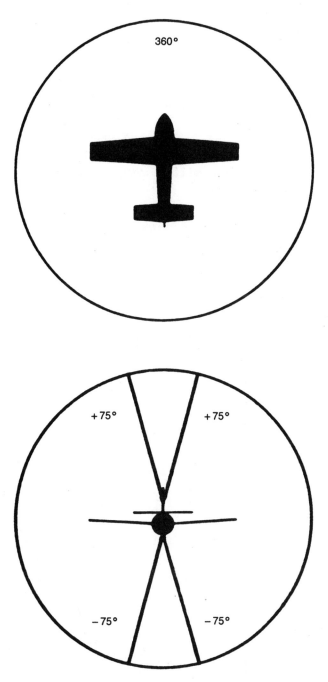

Fig. 1-4. *FAR 23.1401 strobe light projection requirements.*

so great it can easily cause short-term blindness of pilots in nearby aircraft. And there is still a problem with strobes reflecting off of airframe components, especially off the back of props. If you have strobes on your aircraft, it is a good idea to paint the back of propellers, nacelles, and wing surfaces with nonreflective black paint. If you do this, check it on the ground, at night, the first time you fly the airplane to be sure there are no surprises in flight. Strobes present still more potential problems.

Some systems cause communication and navigation radio interference. The problem is strobe lights function as a result of capacitor discharge, which can cause radio frequency interference (RFI), especially in low-frequency equipment such as ADF and LORAN C. It can also cause an audible clicking sound in the speaker and headphones. None of this affects the signal, but it can be very annoying to listen to it for an extended period of time. If your aircraft suffers from this malady, it is almost always the result of poor installation technique. Typically it can be reduced, and even eliminated, with cable shielding. Don't confuse that problem with the normal "wheeeep...pop" tone of the strobes' power supply, which is audible when the engine is shut down; that's a normal condition.

As always, an ounce of prevention is worth a pound of cure. You're better off paying a little more and having a strobe system installed by someone with a lot of experience. They will make sure never to locate a strobe power supply closer than 3 feet from any antenna, especially a low-frequency one. Also, the lamp-unit wiring should never cross or lay next to nav/comm wiring; bundling wires is a major cause of interference. And something as simple as ensuring that the power supply case and shielding around the wires are properly grounded to the airframe makes a major difference.

Other Required Lighting

The only other exterior aircraft light required by regulations is an icing light for aircraft certified for flight into known or forecast icing conditions. It is an incandescent light installed in the fuselage or on an engine nacelle, which illuminates the wing's leading edge to monitor icing in the dark. Contrary to popular belief, landing lights are not required unless the aircraft is operated for hire at night either VFR or IFR.

Lighting Preventive Maintenance

There are three types of position light bulb replacements available. The first is a 21-watt bulb, which simply won't meet minimum intensity requirements at typical line voltage and should not be used. The second is a 26-watt bulb, which requires 13.2 volts at the light assembly to comply with the FAR intensity requirements. Unfortunately, this does not allow for plastic enclosures, such as an enclosed wingtip. Plastic enclosures over position lights alter the light's output from 5% to 30%, depending on the clarity and the radius of the curve through which the light is projected, making it all too easy to drop below minimum intensity requirements. The third is a 37-watt bulb that will comply with the 12.5-volt system capability and provide a much brighter light than required at normal line voltage.

Keeping a strobe light system healthy is a bit more complex than simply changing bulbs occasionally. The power supply's longevity depends on regular use of the system. Extended periods of nonuse result in the electrolytic condenser losing polarity formation. If a system is not used for over a year, there is significant potential for the system to fail. Should you find yourself in this situation, the solution is to remove the power supply, operate it at 75% of the normal voltage for 10–15 minutes, then reinstall it and use the system normally. If you remove a power supply, remember that these are condenser-based systems. Use caution, as the condenser can build up and store approximately 450 volts DC.

To periodically check the health of the xenon flashtube, you can use a paper cup like a stethoscope to listen to it flash. A healthy power supply and trigger transformer will make a "snapping" sound much like a spark plug. You can also feel the snap, which is a very low energy pulse, by placing a finger in close proximity to the strobe light. Xenon tubes are highly photosensitive, so a tube going bad may fire in the daylight because the ambient light conditions aid it, but the same tube may not fire at night. The tube may also fire when the engine is running but not when it is shut down because of the lower battery output.

Flash tubes are also subject to the effect of aging and will prematurely age when kept routinely in areas of high temperatures. The tube itself is airtight, and occasionally a leak will develop as a result of eggshelling of the glass or partial failure of the seals where the wire enters the glass. This condition is inevitable, as it is caused by the hot and cold cycling of the system; everything gets old and wears out.

One common reason for replacement is self-ionization. When one or more strobes begin to glow a continuous light blue, the entire system will become inoperative. This typically will occur when the system voltage is highest. To verify that the problem is ionization, turn off the system, wait a few minutes, then turn it back on. If everything operates normally for a minute or so, then fails and you see the glow, you've got self-ionization; replace the affected xenon tubes.

When replacing an old xenon tube, you may discover that other tubes begin to misfire or skip. In most cases, this signifies that they too are getting near the end of their service life. To check them out, remove the tube and operate it at 20% of its normal input power supply. If it operates at the reduced level, it still has good service life left. Another problem, found only in double-flash systems, is an intermittent second flash. Such systems are designed to operate at normal line voltage. What probably is happening is the system voltage is dropping below the battery-charging voltage and the second flash isn't getting off. Other electrical system problems probably are to blame rather than the strobe system.

CONDUCTING AN INTERIOR PREFLIGHT

The most important aspect of an interior preflight is carefully adhering to the published checklist. Beyond that, check the cabin for cleanliness. Dust floating in the air during flight can cause eye and respiratory problems. As a student pilot, I found that the airplanes were so dirty at my local FBO that for a long time I logged two sneezes for every touch and go. Loose objects are also a major concern in an aircraft cabin. In turbulence or during a fast stop, they can become lethal projectiles hurtling through the air.

Chapter One

Always check the security of the seats and particularly the pilot's seat. There have been several fatal accidents over the years as the result of a pilot's seat sliding back on its tracks just after takeoff. Also check the integrity of window and door locks. While windows popping open in flight tend to be more frightening to passengers than anything serious, a door popping open can have consequences. Most aircraft can fly with an unlatched door, but I once had a door pop open during an instrument approach in a multi-engine aircraft and it scattered the approach charts, in my admittedly open binder, all over the aircraft. Baggage and access doors, on the other hand, can cause problems if they open in flight, depending on where they're located. Twins with wing lockers may have baggage come streaming out. Single-engine aircraft cowling doors can tear off and fly into the windshield.

Finally, an aircraft with an organized cabin is easier and less stressful to fly. Have your pilot's bag behind the front passenger's seat so it is easy to reach across and find something. Put things in the same logical places all the time so you know where to reach for them when the pressure is on. And make sure they're secured so they don't float around in flight unless, of course, you want to impress your friends.

When I was a young, brash flight instructor, I used to call my handmade IFR clipboard George and always leave it right on the instrument panel when flying with a student. Inevitably, students would ask why I called it George and I'd tell them that I had designed and built it to do most of the organization for me so the old saying about "Let George do it" had come to mind. Then I would tell them George and I had bonded and I wouldn't fly without him. In fact, George will even come to me if I call him, I would tell them with a deadpan expression.

Now flight students can be naive and they certainly don't want to get on the wrong side of their instructor, but that was too tall a tale for anyone to take quietly. So they'd challenge me, we'd bet a cup of coffee on it, and then I'd simply extend my hand in the air above it, call out, "Come, George," and pull a few negative Gs so it would float up and into my hand. Okay, it was kind of a stupid trick in retrospect, as if a trainer aircraft needed any additional Gs, but it never failed to get a laugh from the student and a free cup of coffee for me. That little game ended when one day a student surprised me and stalled out the aircraft on takeoff. As I was wrestling the controls from the student, George kissed me right between the eyes and came close to getting all three of us killed. From that day forward, everything was secured in the cabin of any aircraft I flew.

TRANSPARENCIES

For all the years I taught in collegiate aviation, I don't think I ever encountered a student who knew that aircraft windows are properly called transparencies. For many years transparencies were exclusively made of glass. There is actually a lot to be said for glass, as it has excellent optical qualities, is easy to maintain, and is very resistant to scratching. In fact, scratches were often the result of windshield wipers dragging trapped sand and dirt across the surface of the glass, a problem that could be virtually eliminated by cleaning the blades often and never running the wipers when the window was dry. Light to moderate surface scratches in glass are acceptable until they get to the point that they cause a visibility problem. Even then, light to moderate scratches are usually removable by someone with experience, although if done incorrectly, the result will be irregular glass removal and distortion. Glass does have, of course, some drawbacks.

Glass is very difficult to shape, so flight deck windshields tended to be composed of a series of fairly short, flat sections. Glass is also expensive and is very heavy compared to the primary alternative: acrylic transparencies.

Most pilots use the term Plexiglas, which is actually the trademark for one type of single-ply acrylic manufactured by the Rohm and Haas Company of Philadelphia. The Plexiglas formula was patented during World War II when U.S. fighter aircraft needed contoured canopies. The correct generic name is actually "as-cast" acrylic transparency.

Made of a monolithic polymer, acrylic molecular structure is uniform throughout, so it does not have flaws or discontinuities, and acrylic is molded easily to the contours required by modern aircraft. Even though there is no strengthening treatment, as-cast acrylic is remarkably strong and resilient. It actually has a greater impact resistance than glass.

Scratches in acrylic transparencies usually don't require replacement unless they interfere with vision. Since windshield wipers are not put on acrylic transparencies, the cause of scratches is almost always physical abuse. There are three types of scratches that can be removed relatively easily with the proper equipment and knowledge. The first is the hairline scratch, usually caused by improper cleaning procedures. The second type is the minor scratch that you can feel with a fingernail. The deep scratch you can feel with the tip of your finger is the third type. Cracks and crazing do not fall into any of these categories.

Most light aircraft use single-ply as-cast acrylic transparencies, which eliminates the problem of flaking or delamination—when one ply separates from another. As-cast transparencies are also relatively inexpensive and hold up fairly well for several years in the outdoors before deterioration and discoloration set in. Acrylic can also handle significant temperature changes and is designed to absorb ultraviolet rays, which is especially important in preventing pilot sunburn at higher altitudes.

When it comes to noise suppression, however, the typical light aircraft doesn't do so well, as it has an acrylic transparency ranging from .19 to .25 inches thick. The noise from the props, engine, and wind can be remarkably loud, though it is possible to get thicker transparencies with a bit more sound-deadening capability on the after-market. There are, of course, problems even with as-cast acrylic transparencies.

The biggest disadvantage is that acrylics lack hardness. With approximately the same surface hardness as brass, they are easily scratched. Worse, many cleaning agents and solvents attack, and some literally eat, acrylics. You have to keep a very close eye on who washes your windows.

Stretched acrylic is stronger than as-cast acrylic and as a result is used on many pressurized and some large unpressurized aircraft. The process stretches a sheet of thick acrylic into a thinner, specified shape. The act of stretching makes acrylic more resistant to cracks and less susceptible to crazing and abrasion. *Crazing* refers to those tiny surface cracks that, when the light hits them from certain angles, turn the windshield into a blinding glare. Stretched acrylic is engineered to withstand greater stresses, temperature extremes, pressurization stresses, and in-flight hail encounters. In fact, they are manufactured to be able to handle bird strikes as high as 360 knots!

There is actually a test to determine bird strike compliance. A frozen chicken of a specified weight is loaded into a specially designed cannon and shot at the windshield to simulate the bird strike. I know that sounds like a joke, but that's actually how the test is conducted. I think the inventor of the process was awarded a Pullet Surprise for his pioneering efforts.

Laminated acrylic is a layer of vinyl sandwiched between two layers, or plies, of stretched acrylic. The vinyl commonly contains a heating element of fine mesh wire used for windshield deicing, which has unfortunately been proven to cause a few problems. While it is great for windshield deicing, it turns out that acrylic is sensitive to high temperatures. Some aircraft transparencies have crazed and even cracked as a result of excessive heat from the deicing system. So, on some large twins, the captain's side of the windshield is made of glass, which, because it is less sensitive to temperature, is equipped with an electric deicing system. The copilot's side is made of a less expensive, unheated acrylic transparency.

Most single-engine and light twin-engine aircraft use as-cast acrylic resin transparencies. They actually transmit light better than glass—about 90% of the available visible light through untinted windows. Of course, there are still the poor hardness qualities to contend with, making them very susceptible to scratches, distracting reflections, and severe glare problems.

Care of Acrylic Transparencies

Proper care of transparencies is essential to prevent problems, ensure a long life, and reduce potential glare problems. First, never clean a window with a coarse cloth or paper towel. You really have to keep your eye on those "efficient" flight line attendants. They also seem to particularly enjoy carrying around a bottle of chemical cleaner that will react with and damage acrylic. In the long run, you are always better off cleaning your own windshield to be sure the job is done properly.

To clean glass transparencies, you should first remove excessive dirt with clean, flowing water. Then clean the transparency with a solution of mild detergent, such as Joy or Ivory Liquid, and water—or, if you prefer, with a 50% solution of isopropanol and water; either will do a good job. After applying the solution with a soft, clean cloth, rinse the area thoroughly and then dry it. The use of window cleaners, such as Glass Wax or Windex, is not recommended because even though they clean well, they contain wax which can cause streaking.

The proper method for cleaning acrylic transparencies begins with your removing your rings, watches, and other jewelry, which can cause deep scars in the surface. Next, remove excess dirt from the transparency with a flow of clean water. You can locate and remove caked dirt with your hand and fingers while flooding the area with water. Thoroughly clean the crevices around the window framework; dirt hidden there may be dragged out later and scratch the surface.

Next, wipe the surface with a clean, soft cloth or sponge using a warm cleaning solution of isopropanol to remove grease or oil on the transparency. You may want to experiment with different cleaning agents to see what works best for you. If so, try aliphatic naptha type 2, hexane, or kerosene, but when using the latter two be sure to flush the area with water when the grease or oil is removed. Never use any abrasive materials, strong acids or bases, methanol, methyl ethyl ketone (MEK), or any ammonia-based glass cleaner, as all of them will damage the acrylic. And despite their ready availability, be careful about using approved plastic cleaners in aerosol cans. The chemical itself may be both safe and a good cleaning agent, but the aerosol propellant chemicals may damage the acrylic. It isn't like cleaning your car, is it?

Here's something else to consider; choose a cleaning rag carefully. Any of the following are considered safe: 100% cotton flannel, 100% cotton terrycloth, or genuine chamois, but not the synthetic or imitation kind. These rags may be reused if cleaned thoroughly, dried, and stored in plastic bags to prevent contamination from dust. Always avoid using paper towels unless they literally feel like cloth when rubbed against your face. When in doubt, don't use it. And certainly avoid using shop towels, even clean ones. Shop towels are used to clean up metal shavings and all kinds of chemical spills; you never know what is trapped in the fibers.

When cleaning a transparency, never rub in a circular motion, as it causes "glare rings." Instead, you should rub in one direction, preferable up-and-down, as horizontal scratches are more noticeable than vertical. It is also a bad idea to rub more than once with the same section of cloth because the dirt on it will scratch the transparency. Take a good look at the cloth after you make a pass over the windshield with it; you have just created homemade sandpaper! Instead, fold the cloth so as to expose a clean area and wipe again. When finished, rinse thoroughly, dry the cloth, and store in a plastic bag.

You should also avoid excessive rubbing with a dry cloth, as it scratches the surface and builds up a static charge that attracts dust particles. When you're done washing the window, remove excess water with a chamois, but don't completely dry the transparency with it. Instead, let it air dry. Once completely dry, many owners will polish the transparency with a thin coat of hard wax such as Johnson's J-Wax or Turtle Wax. While that's a good idea in principle, in practice there's a problem. Many automotive waxes contain abrasives that will scratch the surface of acrylic. It is a better idea to polish the transparency with Micro-Surface Finishing Products—Micro-Mesh antistatic cream, which is designed to remove water spotting and leaves a high-gloss protective finish on the surface. Waxing the transparency prevents pitting by reducing its water-absorption capability.

The Three Transparency Killers
There are three naturally occurring killers for acrylic transparencies: ultraviolet light, water, and abrasion.

When aircraft are stored in outside tiedowns, the acrylic transparencies are left exposed to the sunlight all the time during daylight hours. You can count on a useful life of approximately two to three years! Ultraviolet breaks down the composition of plastic, causing acrylic transparencies to age and get brittle. The telltale sign is a yellowish color permeating the entire thickness, which cannot be removed. New transparencies are designed to bounce a small bird in flight, so if you gently push on a new window it should flex slightly. An old window, especially one that has spent its life in the sun, could easily crack by doing the same thing. It is always the best course of action, and less expensive in the long run for many reasons, to store an aircraft indoors.

The second naturally occurring killer is water. Acrylic is hygroscopic, meaning it absorbs water. Transparency thickness fluctuates slightly all day long as it absorbs from 2 to 10% of its weight in water, depending on the humidity. In flight, the aircraft will naturally build up a negative static charge, causing the water to act as a discharge wick. Static electricity literally explodes from the window, leaving a permanent pit behind. The best preventive maintenance is a half-hour per week spent waxing transparencies with paste wax to prevent them from absorbing the water.

Finally, there is abrasion. Your friendly flight line attendant is your worst enemy. While wiping your transparencies with a rag previously used to mop up metal shavings would be bad enough, even more commonly it is the liquids that may be absorbed in the rag. Solvents, paint strippers, acetone, and especially hydraulic fluid literally destroy acrylic transparencies. One FBO, mindful of this problem, issued a directive that personnel would not use rags of any type to clean windshields; rather, they would use new paper towels of the type found in the restroom. Unfortunately, the towels were rough enough to cause damage by themselves. They also set up a terrific static charge that strongly attracted dust and other airborne particulate matter, probably causing more damage than the towel's roughness. It is always best to take the time to clean windows yourself. For the same reasons, it is good preventive maintenance to keep the cockpit as clean as possible. A clean cockpit cuts down on airborne dust, which, after adhering to the inside of the window, will eventually scar it.

Acrylic Transparency Replacement

A glass transparency must be replaced immediately if it is cracked, but that is not necessarily so with acrylic. One of the advantages of an acrylic transparency is that if a crack in the transparency of a nonpressurized aircraft is caught early enough, it is possible to stop-drill it to prevent further damage. Cracks are often the result of relief of internal stresses caused by the manufacturing process or installation. While stop-drilling a crack is certainly cost-effective, pilots should be aware that this procedure does weaken the structure somewhat and the transparency becomes less likely to be able to hold together in the event of a bird strike. Also, the potential for in-flight structural failure may also be somewhat greater. A crack in the outer ply of a laminated acrylic transparency weakens the entire structure somewhat, but a strong inner ply is still capable of withstanding normal stresses. On the other hand, a crack in the inner ply significantly weakens the structure; such a transparency should be replaced.

Delamination

One of the more serious problems associated with transparencies, be it glass or acrylic, happens when the inner layer separates from the other layers. Called *delamination*, it is typically caused by deterioration of the weather sealant around the transparency. Water seeping past the sealant gets between the plies and the area takes on a cloudy or milky appearance. When this happens the entire transparency should be replaced.

Initially, delamination may only be a problem of visibility. Moderate delamination does not particularly affect the bending or tension capability of the windshield, so if the visibility is good enough, there is no urgent problem. Some delamination along the edges of the transparency is normal, as window edges have limited adhesion to allow for temperature expansion. With glass windows, if delamination expands into the window and develops an irregular or jagged boundary, that is an indication of a lack of uniformity of the separation. This situation causes the polyvinyl butyral inner layer to pull chips of glass from the inner glass surface, which will cause failure of the glass ply. If chipping is present, or if the problem becomes worse, the transparency must be replaced. Smooth-edged, clear delamination typically does not get worse, as it indicates that the original stress causing it has been relieved.

Crazing

Probably the most common transparency-related problem is the network of fine cracks, which appears to extend all over the surface. Crazing does weaken the strength of the transparency.

The individual cracks are difficult to discern, and you have to view them by looking approximately perpendicular to the transparency's surface. The cracks are narrow in width, typically not more than about 0.001 inch deep, and usually lie below the surface rather than on it. For that reason it is usually impossible to feel them with a fingernail. There are two categories of crazing: minor and severe.

With minor crazing you typically don't see the lines themselves because they are so small. Instead, you see a distinct discoloration or milky appearance reflected in the light. Severe crazing may actually have fewer scratches, but they are significantly larger, appearing to be deep gouges in the surface. Again, these are actually under the surface, and while they can be removed with heavy but even sanding, it will require the removal of a substantial amount of acrylic material to clean up the transparency, after which an industrial repolishing procedure is required.

As I walk around airports, I see a lot of owners use external window covers. When I inquire about them, I'm always told they are there to save the windows and block the sunlight from overheating the cabin and potentially frying the radios. The real value of such covers is debatable. It's a feel-good solution that may have more of a downside than an upside. While they will be useful inside a hangar to protect the transparency from birds, outside they can be downright destructive to your airplane as they flap wildly, beating dirt against the window and accelerating deterioration. If cabin heat is a real concern, use internal Velcro-mounted covers instead. They will act as a heat shield and security device because they prevent anyone from looking inside the cabin and seeing the avionics and other installed equipment. However, even they may still seriously raise the temperature of the heat-sensitive acrylic, which can cause crazing and, in laminated windows, bubbles. The best strategy is to hangar an aircraft and avoid using any window covers.

One last thought: when the aircraft is being repainted, great care should be taken to protect the transparencies and sealant. Use high-quality masking material to assure that no paint product comes into contact with the transparency at any time. An open can of paint stripper should not even be put in proximity to an aircraft transparency. Acrylic absorbs stripper fumes, resulting in rapid crazing.

Window Restoration

The restoration of crazed transparencies is definitely possible and even can be done by the pilot, but it is not your typical home project. You must know the panel's exact thickness, as the existing thickness must be compared to the minimum allowable thickness, which varies from manufacturer to manufacturer. The kit will come with an appropriate technical manual that will help you make the determination. The restoration process removes some of the windshield, and it is possible that your transparencies are already too thin for refinishing. Ultrasonic measurement is the best method for the novice, but not many novices have access to ultrasound equipment. After completion, the transparency must again be checked for thickness and compared to assure compliance with the minimum requirement. Some transparencies may be refinished several times, greatly increasing their useful life. All things considered, unless you have a lot of mechanical experience, this process is best left to the experts.

EMERGENCY LOCATOR TRANSMITTERS

At best, search and rescue (SAR) is a hit-or-miss proposition. It wasn't all that long ago when the search sometimes claimed more lives than the downed aircraft. Rescue was often delayed by poor reporting procedures, and the actual search often had airplanes flying haphazardly over suspected areas, sometimes into each other!

The FAA sought to shorten the time between accident and rescue because many victims survived the crash only to die of exposure and hunger. At one point it was determined that 50% of the people who were successfully rescued were found in the first 12 hours; 25% were found within the next 12 hours. The odds weren't good if you had to wait longer.

On December 29, 1970, an amendment to the Federal Aviation Act of 1958 went into effect. It required most U.S. civil aircraft to have an FAA-approved emergency locator transmitter (ELT) installed and operating.

The exceptions, as they pertain to general aviation, are listed in Federal Aviation Regulation 91.207, and are as follows:

1. Turbojet-powered aircraft.
2. Aircraft while engaged in scheduled flights by scheduled air carriers.
3. Aircraft while engaged in training operations conducted entirely within a 50-nautical mile radius of the airport from which such local flight operations began.
4. Aircraft while engaged in flight operations incident to design and testing.

5. New aircraft while engaged in flight operations incident to their manufacture, preparation, and delivery.

6. Aircraft while engaged in flight operations incident to the aerial application of chemicals and other substances for agricultural purposes.

7. Aircraft certificated by the Administrator for research and development purposes.

8. Aircraft while used for showing compliance with regulations, crew training, exhibition, air racing, or market surveys.

9. Aircraft equipped to carry not more than one person.

10. An aircraft during any period for which the transmitter has been temporarily removed for inspection, repair, modification, or replacement, subject to the following: (i) No person may operate the aircraft unless the aircraft records contain an entry which includes the date of initial removal, the make, model, serial number and reason for removing the transmitter, and a placard is located in view of the pilot to show "ELT not installed." (ii) No person may operate the aircraft more than 90 days after the ELT is initially removed from the aircraft.

The subject of required ELTs was very controversial at the time. Over 90% of the distress signals are erroneous. Eighty percent of them originate at airports and once they are finally found, the searchers typically can't contact the aircraft's owner; sometimes they can't even get into the hangar. One ELT was found, after an extensive search, in the pilot's flight bag, located in his hotel room.

Most ELTs are self-contained, portable, handheld devices that mount in the aircraft. Weight ranges from 1 to 5 pounds with the battery pack installed. Some have remote antennas mounted on the fuselage, some have antennas on the unit itself, and some have both types. When the aircraft is subjected to a G load in excess of the permissible range, such as in a crash, the ELT is activated by a "G" switch, which is essentially a vibration sensor.

There are two principle types of activation methods: rolamite switch and sliding mass. The rolamite switch is a "rolling mass" type of switch. Under the influence of high Gs, the switch mechanically closes the activation switch and the ELT begins to broadcast its warble alert. The sliding mass system is a spring-loaded, instantaneous-type switch. The switch closes the electrical contacts, which then hold themselves closed once contact has been made. In either case, once the ELT is activated it transmits a distinct, "wow-wow" modulated tone signal on 121.5 and in most cases also 243.0 MHz. The catch is someone has to be listening.

Installation Considerations

With hundreds of thousands of ELT units installed in the United States alone, manufacturers have learned a few things about installation. A study of accidents reveals some interesting details about survivability. The tail section remains relatively intact approximately 84% of the time, the nose only 1%, and the cockpit only 2%! Of course, the ELT has a much higher survival rate in any of those areas, but that certainly gives you an idea about the best place to mount it.

There are other considerations, too, such as potential for fire damage. And don't forget that not only must the ELT survive, but the antenna and connecting cable too if it is externally mounted. Therefore, it is recommended that the ELT have an attached and externally mounted antenna and the unit be located as far aft in the empennage as possible. Incidentally, 33% of all crashed aircraft end up inverted, so be sure the system is capable of functioning while inverted!

Common ELT-Related Problems

There have been definite ELT problems over the years. Probably the most common problem is inadvertent activation. The G switch can be activated by excessive vibration such as a hard landing. Where an ELT is installed in the aircraft may also lead to problems. Such common locations as baggage compartment and rear seat tend to make the unit more vulnerable to inadvertent activation. Bouncing bags or stretching arms have often caused ELTs to go off. Make sure the proposed location provides adequate protection but still permits easy access to the unit so it can be checked when necessary.

If accidental activation is a problem, so too is lack of activation at the appropriate time. Always use the proper batteries for replacement. Whenever replacing batteries with anything other than original equipment, contact the ELT manufacturer for approval. It really isn't a sales gimmick to talk you out of Brand X; tests have shown that many substitute batteries won't activate the ELT, especially when they get a little bit older. Don't let advertising or even a TSO number fool you; check with the original manufacturer. What happens is some batteries produce a passivation layer as they age; this is especially true for magnesium-type cells. What can happen is the layer can cause a delay of over one second before the cell achieves rated voltage. ELTs with electronic latching circuitry can be activated by a crash pulse of less than 0.1 second, but if battery voltage is too low, the unit simply won't activate. It is important to point out, however, that modern ELTs with such circuitry probably don't require maximum voltage to activate, but why take chances? The problem can be totally avoided by using alkaline (zinc-manganese dioxide) cell batteries or any battery the manufacturer approves. Incidentally, the reason why you might want magnesium batteries is because they have a longer shelf life and provide sustained high power for a longer period of time, especially in extreme cold.

If it sounds like there is a lot to worry about when considering batteries, you're right. The simple truth is that batteries are the most common cause of ELT problems, and you just can't take good enough care of them. For extended storage, batteries should be refrigerated. Cold slows normal battery deterioration, increasing their shelf life. There are some risks associated with refrigeration, though. When transferring the battery from cold to warm air, there is a chance of getting condensation within the cells or even within the entire battery pack. This leads to corrosion, premature battery depletion, and the formation of conductive paths between cells, which cause shorting and battery failure. Some manufacturers hermetically seal the cells to prevent such a possibility, but it is a good rule of thumb to always bring the battery to room temperature as slowly as possible and completely wipe it dry before installation in the ELT unit.

According to the FARs, batteries used in the emergency locator transmitters required by Federal Aviation Regulations must be replaced (or recharged, if the battery is rechargeable)(1) when the transmitter has been in use for more than one cumulative hour; or (2) when 50% of their useful life (or, for rechargeable batteries, 50% of their useful life of charge), as established by the transmitter manufacturer under its approval.

In a study of nonfunctioning ELTs, it was discovered that 14% of the ELTs examined had outdated batteries, while 6% had batteries that were dead. The importance of good preflight and preventive maintenance cannot be overemphasized. You should routinely have the ELT and battery checked; a 100-hour inspection is a convenient time, but certainly at every annual. Periodically test the unit by listening to 121.5 (and 243 if you have a radio capable of receiving on that frequency) and activating the ELT manually. This is an FAA-sanctioned test provided you only conduct it during the first five minutes of any hour and limit it to a few modulations.

In general, it is a good idea to turn the battery off when the aircraft is not in use, but always remember to include BATTERY ON in your preflight. While you're turning it on during preflight, also check the battery expiration date. If it is not readily visible, write it down on the ELT or next to it where it can be easily checked.

When You're the Target

If you find yourself as the object of an SAR effort, there are several things to keep in mind. After checking to make sure the ELT is alive and well, the first order of business is to take the necessary steps to ensure survival. If the conditions are hostile, protect yourself as necessary from the elements.

Remember that the ELT transmits omnidirectional, roughly in a circular pattern, and search aircraft will be using a "build-and-fade" method to locate your ELT signal, as depicted in Figure 1-5. Signals have been received as far as 100 miles away by aircraft operating at 10,000 feet. Transmission is line of sight and can be blocked if the antenna is under the fuselage or any other metal structure; even rough terrain can play havoc with the transmission signal. Of course, a broken antenna is a serious problem but there have been many cases where once an aircraft gets into the general area it has been able to home in on an ELT without an antenna.

If terrain appears to present a problem in signal broadcast, consider relocating the ELT, perhaps higher up in a tree or on a ridge. Finally, it is important to understand that a downed aircraft is extremely difficult to spot from both the air and ground. There have been many instances where searchers have passed within a quarter mile of a wreck and never seen it; aircraft have flown directly overhead with no results. Take whatever steps are necessary to ensure high visibility. Flashing with mirrors and glass are attention getters, as is a smoky fire.

AIRCRAFT ANTENNAS

One of the features of an aircraft that makes it practical for everyday use is the ability to fly in all kinds of weather and operate in and out of airports all over the world. Neither of

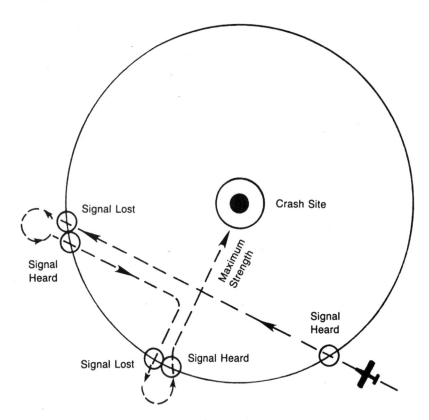

Fig. 1-5. *Build-and-fade method of locating ELT signals.*

these two major issues would be possible if there were no navigation and communication radios. Linking radios to the ground is the role of the antenna.

Usable radio frequencies range from 30 kHz (kilohertz or thousands of cycles per second) to 30,000 MHz (megahertz or millions of cycles per second). The lower the frequency, the greater the distance the signal will travel, but the more susceptible it is to environmental conditions such as lightning. Higher frequencies, while less likely to be bothered by the environment, have the limitation that they require line of sight between the transmitting and receiving antennas. Obviously, altitude becomes important with higher-frequency radios.

Antenna Size and Orientation

Antenna design is a function of the frequency band for which it is to be used. Wavelength is inversely proportional to frequency, so the higher the frequency, the smaller the wavelength and the smaller the antenna. Generally speaking, due to their different purposes, vertically mounted antennas are for communication, and horizontally mounted antennas are for navigation.

Within the radio frequency range, there are several bands designated for specific aircraft use. The low-frequency (LF) band ranges from 30 to 300 kHz. Long-range navigation (LORAN) utilizes the 90–110-kHz range. A LORAN antenna resembles a communications antenna mounted on the top or bottom of the fuselage.

The automatic direction finder (ADF), which operates in the 190–535-kHz range, falls into both the low-frequency range and the medium-frequency range of 300–3000 kHz. The ADF receives signals from the ground-based nondirectional beacon (NDB). Older aircraft used a long wire antenna, usually running from the top of the cockpit to the top of the horizontal stabilizer and a loop antenna on the underside of the fuselage. Newer aircraft have both antenna mounted in a single teardrop or rectangular-shaped box on the underside of the fuselage.

The very high frequency (VHF) range is from 30–300 MHz, and there are several aircraft radios that operate within this range. Marker beacons, at 75 kHz, are mounted on the underside of the fuselage. They look like a thick wire bent sharply back toward the tail or can also be a 6–8" canoe-shaped antenna.

The VOR/LOC (VHF Omni Range/Localizer navigation) operates within 108–117.95 MHz and has the antenna that is most commonly shaped like a 1.5-foot-long "V" that is horizontally mounted on the vertical fin. There are two other antenna configurations that are sometimes seen. One is the balanced loop or "towel rack" type and the other is the combination nav/comm "boomerang" antenna typically mounted on top of the vertical stabilizer.

Communication radios operate in the 118–137.975-MHz range. Each radio has its own antenna, which is a vertical wire approximately 1.5 feet high and is usually mounted on top of the vertical fin or fuselage.

The ELT operates on 121.5 MHz and has a short vertical wire antenna mounted on the vertical fin or fuselage. Some ELTs only have an antenna directly attached to the ELT unit itself, which is located inside the aircraft.

The ultra high frequency (UHF) range is from 300–3000 MHz and supports several aviation radios. The glideslope operates between 329.15–335 MHz, and its antenna comes in several different configurations. It is often in combination with a VOR antenna or located in a wide, oval plastic container at the top and inside the windshield. Sometimes it is a half-foot-wide boomerang mounted on top and outside of the cockpit or perhaps on the underside of the fuselage.

The antenna that supports distance measuring equipment (DME) is a very small, fin-like antenna, as shown in Figure 1-6, and is mounted on the underside of the fuselage. DME operates in the 962–1213 MHz range.

The transponder receives on 1030 MHz but transmits on 1090 MHz. It uses the same type of antenna as the DME or may also use a 2-inch miniature car-type antenna with a small ball on the tip.

Aircraft equipped with a global positioning system (GPS) will show a small rectangular or teardrop fiberglass antenna mounted on top of the fuselage, as it must receive satellite signals, which are broadcast on 1575 MHz.

Fig. 1-6. *DME antenna.* (Photo by author, courtesy of Frasca Air Services)

Preflight Considerations

Preflight considerations are very straightforward. They include security and cleanliness. When looking at individual antennae, check to be sure that the attachment to the airframe is secure, and if connections are visible, check their tightness. Essentially you are checking the general integrity of the antenna itself for cracks, distortion, or other obvious problems.

Dirt, grease, mud, and anything else that can cake on an antenna will degrade performance. This is particularly a problem with belly-mounted antennae, so they should be checked during preflight to assure cleanliness. If they are dirty, simply take a damp rag and wipe them clean. Any abnormal condition should be referred to an avionics technician.

2
Aircraft Instruments

PERHAPS THE MOST FORGOTTEN ITEMS IN AN AIRCRAFT WHEN IT comes to preflight and preventive maintenance, aircraft instruments are the windows on another world. They allow the pilot to "look into" various systems to control them and make judgments about their health, and find a location on the ground when they can't see more than a few feet outside the aircraft. Aircraft instruments are the eyes of the airplane.

PITOT-STATIC SYSTEM

Of the five primary flight instruments (airspeed, altimeter, vertical velocity, compass, and turn indicator), the first three use the pitot-static system. These instruments interpret aircraft performance within an air mass.

Airspeed Indicator

Regardless of the level of sophistication, all airspeed indicators share one thing in common. Airspeed indicators compute dynamic pressure by measuring the difference between air pressure resulting from airplane movement through the air (ram) and ambient air pressure (static). A good example of ram air can be felt when you put your hand out the window of a car traveling at 50 mph. Static air pressure is the ambient atmospheric air pressure that surrounds the aircraft at any given moment.

The airspeed indicator provides some very basic, yet important information on the aircraft's relative velocity through the surrounding air. It helps the pilot establish optimum performance during takeoff and landing; it also shows an impending stall in most aircraft, though the less common angle-of-attack indicator gives a better warning of a stalling condition. Engine and airframe manufacturers use airspeed to establish structural limitations, such as airframe never-exceed speeds, flap and gear operating speeds, and propeller harmonic vibration speeds and RPMs.

Airspeed information is also used for operational calculations such as determining groundspeed and is the controlling factor in optimum climb/descent efficiency for given conditions such as best rate, best angle, minimum controllable airspeed, and glide. In emergency situations, it can be an indicator of airplane attitude when engine power and propeller setting are held constant. In that condition an increasing airspeed indicates descent and a decreasing airspeed indicates a climb. Airspeed is a prime factor in establishing long-range cruise control, including fuel consumption, maximum range, and minimum time enroute.

The construction of the airspeed indicator is simple, as illustrated in Figure 2-1. The otherwise airtight instrument case has a static air pressure vent. A sealed diaphragm is inside the case; the bottom of it is permanently attached to the case while the top is free to move. The diaphragm receives ram-air pressure via the pitot tube and expands or contracts depending on the static pressure and ram pressure differential. The top of the diaphragm is connected by linkage to the indicator needle. The face of the instrument is typically calibrated in knots, though some smaller, older aircraft may display miles per hour, which often is used in advertising to make the airplane speed look better. (One knot means one nautical mile per hour; it is redundant to say "knots per hour." There are 6080.27 feet per nautical mile versus 5280 feet per statute mile; one knot equals 1.151553 statute miles.)

Not much goes wrong with an airspeed indicator, except physical damage to the system and icing. If the pitot tube is blocked by ice, bugs, or trapped water, the airspeed indicator functions like an altimeter. The ram air now trapped within the diaphragm becomes a fixed, static pressure. The static air continues to flow into the instrument case, but its density varies with the aircraft altitude. As the aircraft climbs, density decreases, causing the static air trapped in the pitot tube to expand the diaphragm and cause an indication of increasing airspeed. Similarly, a descent would cause the density of the air blocked in the pitot tube to increase, leading to an indication of an apparent decrease in airspeed.

A blocked pitot tube is a dangerous situation; picture a climbout in instrument conditions. Apparently your airspeed is a little fast, so you instinctively ease the nose up just a bit without watching for the results. You continue climbing and again notice your airspeed is a little bit fast, so you ease the nose up more. This time you get an unexpected stall. It is important to understand that when the pitot tube is blocked, the airspeed indicator no longer measures airspeed.

If the static port is blocked but the pitot tube remains clear, the airspeed indicator becomes an altimeter in reverse, showing an airspeed increase with decreasing altitude. When in areas of potential icing, airspeed corrections should be observed whenever a

KNOW YOUR PLANE'S PITOT-STATIC SYSTEM

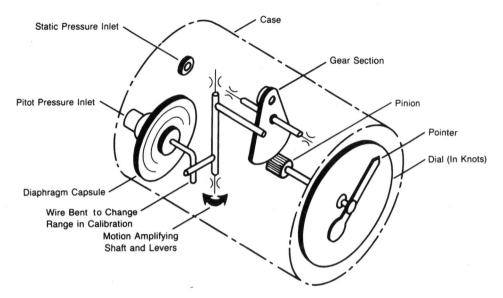

Fig. 2-1. *Simple airspeed indicator.*

change of pitch or power is implemented. If the problem is the result of pitot-static blockage, the error will continue to increase without any sign of the correction having been made. Similarly, always maintain a complete instrument scan; all instruments should agree that there is an airspeed problem; otherwise there is an indicator problem.

The airspeed indicator is designed to be most accurate at sea level on a standard 59-degree Fahrenheit day. At best, indicated airspeed is fraught with errors. It is subjected to various system errors, leaks, and problems associated with improper pitot tube and static source placement. To compensate, the airframe manufacturers give us calibrated airspeed (CAS), which is corrected for system errors. Appropriate airspeed corrections are presented in the pilot's operating handbook (POH) for different conditions of flight—most commonly gear and/or flaps up and down.

Because we seldom operate at sea level, and occasionally fly higher-speed aircraft, pilots must also understand the effect of compressibility on CAS. As density decreases, compressibility error increases. At approximately 10,000 feet an aircraft flying at 250 knots or faster begins to display a serious airspeed error. At that point it is necessary to correct CAS for density variation to compute equivalent airspeed (EAS).

High-speed, high-altitude airplanes encounter such high-ram air pressures that the "ram effect" causes the air to compress, affecting the airspeed indication. This is a well-known phenomenon that affects all aircraft in the same way. Charts provide compressibility factors based on altitude and CAS. Incidentally, this compressibility error applies only to the standard airspeed indicator. True airspeed (TAS) indicators and Machmeters automatically correct for air compression. All aircraft are affected by decreased atmospheric density or what pilots call high-density altitude. The less dense the air, whether the result of high altitude or hot temperature, the greater the error between CAS and TAS.

A calculation must be performed that takes temperature and pressure into consideration to convert CAS (or EAS if applicable) to TAS. Some pilots buy true airspeed indicators, which do the conversion automatically; others buy a less expensive basic airspeed indicator with a rotating dial around the outside, as shown in Figure 2-2. The dial, when manually set to the correct pressure altitude and outside air temperature, gives the pilot the TAS conversion. If you do not have an airspeed indicator that shows TAS, it is possible to calculate a reasonable estimate by adding 2 percent of the indicated airspeed for every 1000 feet of altitude. For instance, if your indicated airspeed at 5000 feet is 100 knots, your TAS would be approximately 110 knots. When you take TAS and correct it for the ambient wind, the result is your ground speed.

Sensitive Altimeter

To the untrained individual, the altimeter can be a very misleading instrument. The altimeter measures the weight of the air above the aircraft rather than the distance between the airplane and the ground. While that may appear illogical, consider what would happen if the altimeter actually measured the distance between the airplane and the ground. Imagine trying to fly at a constant altitude; the indicator needle would bounce up and down every time it passed over a hill, rock, or building. Instead, we fly at a constant pressure

Fig. 2-2. *Airspeed indicator with true airspeed dial.* (Photo by author, courtesy of Frasca Air Services)

level using an altimeter that we have adjusted manually to closely approximate sea level. This adjustment, known as an altimeter setting, already has been corrected for sea level.

Airport elevation is given in feet above sea level. The trick is for the pilot to get an accurate altimeter setting from someone on the ground at the destination airport. If someone at the airport issues the correct information from an accurate instrument, the pilot puts in the correct altimeter setting. If there are no fast-moving cold fronts in the area to play havoc with the pressure, the altimeter should show airport elevation as the airplane touches down.

Older altimeters had one diaphragm (aneroid) and one hand. Their entire range was perhaps only two revolutions of the instrument, and they were not very accurate. Modern altimeters, known as "sensitive altimeters," have two or three aneroids, as depicted in Figure 2-3. If you hold an altimeter over your head and read it, then put it on the floor and read it again, a person of average height will notice that the longest hand indicates a change of approximately five feet.

Of the three major types of altimeters—three-pointer, drum pointer, and counter pointer—the most common in light, general-aviation aircraft is the three-pointer (Figure 2-4, Figure 2-5, and Figure 2-6). On this instrument the longest hand registers 1000 feet per revolution, with each number around the dial equaling 100-foot increments. The wider but shorter hand registers 10,000 feet in one revolution, and each number signifies 1000 feet. The smallest hand would register 100,000 feet if it ever made a complete revolution; each number it points to is read × 10,000.

The one thing all altimeters have in common is their ability to be misread. Numerous accidents have resulted from altimeter misreading, particularly the 10,000-foot indicator. More than one pilot has been cleared to an altitude such as 12,000 feet, misread the altimeter by 10,000 feet, and ended up at 2000 instead!

Because the altimeter must be compensated for nonstandard atmospheric pressure, if it is to indicate the aircraft's true altitude, there must be some way to adjust it (temperature variation is automatically compensated for with an internal bimetallic strip). The pilot dials in the local altimeter setting, which adjusts the drive mechanism within the altimeter to compensate for nonstandard conditions. An entire generation of pilots has called this the Kollsman Window without knowing why; it was the Kollsman Instrument Company that invented the process. In fact, Kollsman is the unofficial granddaddy of altimeters, having invented the first reliable altimeter in 1928. All modern altimeters now have barometric pressure adjustments.

When the aircraft is at sea level in standard conditions (59 degrees Fahrenheit, 29.92 in. Hg.) the altimeter should read zero feet if it is set at 29.92 in. Hg. If the pressure drops to 29.42 in. Hg. and the pilot does not change the altimeter setting, the altimeter will indicate a slow climb to 500 feet ($29.92 - 29.42 = .50 \times 10$). A change of .01 in. Hg. = 10 feet; .10 in. Hg = 100 feet; 1.00 in. Hg. = 1,000 feet. Therefore, the pilot always should keep the altimeter set to a current source within 200 miles of the aircraft's present position and always update the altimeter for each point of intended landing.

The altimeter setting, though often called barometric pressure, is not the same thing available from a local weather station's barometer. Despite the same scale (in. Hg.) and occasionally similar readings, they are in fact different, and only an aviation altimeter setting should be used.

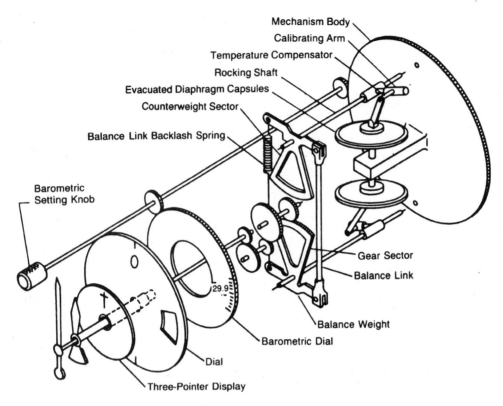

Mechanism Body
Calibrating Arm
Temperature Compensator
Rocking Shaft
Evacuated Diaphragm Capsules
Counterweight Sector
Balance Link Backlash Spring
Barometric
Setting Knob
Gear Sector
Balance Link
Balance Weight
Barometric Dial
Dial
Three-Pointer Display

Fig. 2-3. *Modern, two-aneroid altimeter.*

Fig. 2-4. *Three-pointer altimeter.*

There are four basic types of altimeter errors: scale, friction, mechanical, and hysteresis. Scale error is the result of the aneroids not responding uniformly to the local pressure difference. This type of error is irregular throughout the instrument's range and difficult to predict. A greater margin for altimeter error must be taken into account during high-altitude operations.

Friction error is inherent whenever there are moving parts. Most commonly, the 100-foot pointer will "hang up." This tends to be less of a problem in reciprocating-engine aircraft than in jets, because they vibrate sufficiently to prevent the pointer from sticking. Jets, which have smoother running engines and minimal vibration, use instrument panel vibrators to help overcome friction error.

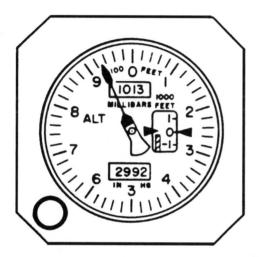

Fig. 2-5. *Drum-pointer altimeter.*

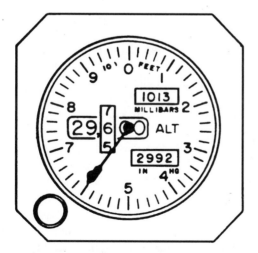

Fig. 2-6. *Counterpointer altimeter.*

If you are flying straight and level for an extended period of time and it appears as if you are doing an exceptional job of holding altitude, it is a good rule of thumb to gently tap the instrument; you may be in for a surprise! If you know the altimeter has "hung up" but a gentle tap doesn't release the pointer, simply twist the altimeter setting knob. Hard-to-twist knobs may indicate internal instrument corrosion, probably the result of static source moisture. If the pointer still doesn't move, static-source blockage, especially icing, would be the next logical suspect.

Mechanical error would be the result of misalignment of gears or slippage of gears or linkage. For the most part, if the altimeter gives accurate indications on the ground during preflight, it probably will be mechanically sound during flight. Actual in-flight mechanical failure is rare, provided the pilot follows up on problems found during pre-flight. Occasionally the setscrew in one of the altimeter's hands will work loose and the needle will drop straight down. If it is the long needle (one revolution is 1000 feet), the medium needle will provide a reasonable indication of altitude within 200 feet. It is not a good idea to be flying low approaches in this situation. If the medium-length fat needle works loose, you only need to keep track of the revolutions of the long needle to assure knowledge of altitude in thousands of feet. In case of a complete altimeter failure, there are two ways of getting a rough idea of your altitude.

If you have a constant-speed prop, you can set the friction lock on the throttle and prop so they won't accidentally move, and use the manifold pressure gauge as a crude altimeter. That's about as rough as crossing the Atlantic in a life raft, but any port in a storm... In a pressurized aircraft, you can depressurize the cabin and use the cabin pressure indicator for a fairly effective altimeter.

Hysteresis is a result of the elastic quality of the aneroids. After maintaining a constant altitude for an extended time, the aneroids require a little time to respond to quick changes in altitude. While it was never a very significant problem, it has essentially been eliminated in modern altimeters.

Absolute altitude is the altitude above the terrain directly below the aircraft. Pressure altitude is for reference purposes and is the altitude above a standard datum plane (sea level, standard).

Density altitude is pressure altitude corrected for temperature and will be the same as pressure altitude only when ambient conditions are standard. Indicated altitude is whatever is displayed on the altimeter, while calibrated altitude is indicated altitude corrected for installation error. Then we have true altitude, which is calibrated altitude corrected for nonstandard atmospheric conditions; theoretically, it is your actual height above mean sea level.

Confused yet? There's one more: flight level is an altitude of constant atmospheric pressure, which relates to the standard datum plane. All aircraft at or above 18,000 feet fly flight levels by adjusting their altimeters to 29.92 in. Hg. The practical application of all this isn't that difficult to understand. When flying below 18,000 feet, use the current altimeter setting from a reliable source within 100 miles of your location. When flying at high altitude, you simply set your altimeter to 29.92 in. Hg. as you pass through 18,000 feet. One last note: If you are going to fly IFR, both the altimeter and static system must have been inspected within the preceding 24 months.

Vertical Velocity Indicator

The purpose of the vertical-velocity indicator (VVI) is to indicate rate of change in altitude by measuring ambient static pressure. The VVI, as shown in Figure 2-7, is calibrated in positive and negative feet per minute and also serves as a reference for level flight.

The VVI is a differential pressure instrument similar to the airspeed indicator. Inside the instrument case there is an aneroid vented to the outside ambient pressure, which operates a needle through a series of gears and levers shown in Figure 2-8. Instead of using ram-air pressure, the case is vented to static air that passes through a calibrated restrictor. As the altitude changes, the diaphragm responds immediately to the free-flowing static pressure. The case pressure, through the restrictor, has an inherent airflow lag, which artificially creates a pressure differential. Manufacturers have developed mechanical compensations for the retarding effect of cold temperatures on the airflow through the restrictor, and to the inherent lag while differential pressure builds. Otherwise the VVI has changed little over the years. Used as a supporting trend instrument, the VVI will help the pilot establish constant rates of climb and descent.

Pitot Tube

Named for Henri Pitot (sounds like pea-toe), the French engineer who invented it, the pitot tube measures the flow of fluid. Adapted for use in airplanes, the pitot tube measures dynamic (ram) air pressure. Pitot tube placement is carefully considered by the manufacturer and becomes more critical as the normal operating speed of the aircraft increases. At subsonic speed the total pressure error as a result of a less-than-optimum position usually will be negligible if the pitot tube is not located in a wake, boundary layer, or region of supersonic flow. An extended boom in front of the fuselage probably is the most effective area, but is difficult for normal ground operations.

Typically, the pitot tube on larger aircraft is located on or near the nose, but light, general-aviation aircraft will generally have it mounted under the wing. Often a small drain hole is added to permit rain to pass through, as illustrated in Figure 2-9. Sometimes

Fig. 2-7. *Vertical speed indicator.*

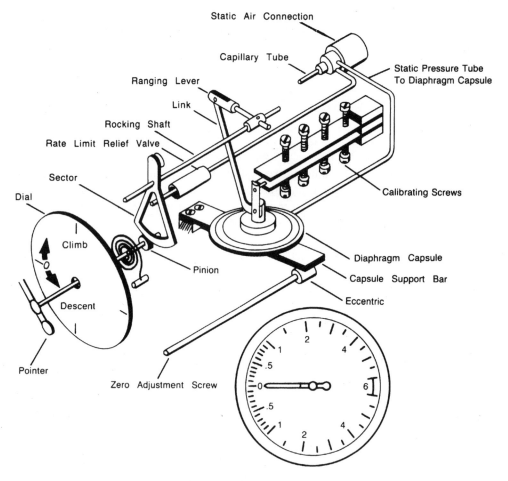

Fig. 2-8. *Schematic of a vertical velocity indicator.*

a sheet-metal vane is added to keep insects out of the tube when the aircraft is on the ground. In flight, airflow swings it open, while on the ground, gravity closes it.

Pitot tubes range in type from the simple bent piece of tubing described to a pitot head with static source and heat, as shown in Figure 2-10. Pitot anti-icing, accomplished by electrically induced heat around the tube, was developed in the 1940s by Aero Instruments for the U.S. Navy.

The most common cause of pitot failure is actual physical damage to the pitot tube itself. The next most common cause is heater burnout, primarily the result of using pitot heat on the ground. Third would be burning a pitot cover onto the tube, which is a particular problem with pitot-static combination probes because the plastic can literally melt into the static port.

Functionally, ram air enters the pitot tube and is brought to a complete stop, allowing pressure to build up to total free-stream pressure or "head" pressure. It is then transmitted to the diaphragm inside the airspeed indicator. Head pressure equals dynamic pressure plus static pressure, therefore head pressure minus static pressure equals dynamic pressure (indicated airspeed).

Static Port

The static port senses the ambient atmospheric pressure surrounding the aircraft. Ideally the port is located in the boundary layer of the fuselage. The column of air in the static port should be perpendicular to the local air velocity, which is exactly equal to the free-stream velocity of the airplane. If it is exposed to rushing air—from turbulence, for instance—the resultant ram-air pressure will cause the static pressure to increase and/or vary. Airflow passing by, which has a greater velocity than free stream, will cause lower-than-static

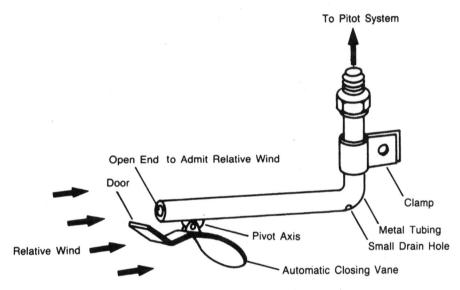

Fig. 2-9. *Pitot tube for a light plane.*

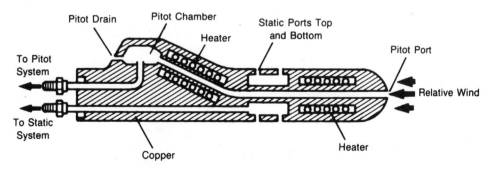

Fig. 2-10. *Heated pitot-static tube.*

pressure. To compensate for problems caused by side-slipping the airplane, it is common for a manufacturer to put two or more static ports on opposite sides of the airplane and cross-couple them to create an "average" static pressure.

The most common source of pitot-static system problems is related to the static source rather than the pitot tube. There are two theories of where to mount a static system port. One is to colocate it on the pitot head, and the other is to mount it on the fuselage. In the fuselage-mounted static source, accuracy decreases as airframe longevity increases because any dings, dents, or abnormalities around the static port changes the boundary layer flow, which in turn affects airspeed and altitude calculation. It is most common for fuselage-mounted static ports to produce errors that vary with the configuration of gear, flaps, angle of attack, airspeed, and even the type of aircraft.

Pitot-mounted static sources are located on a finely machined surface on the pitot probe itself, and the airframe condition hardly affects it. This type of system remains accurate significantly longer than fuselage-mounted static systems. For the light general-aviation aircraft however, the cost of a finely machined pitot-static head may outweigh the benefits. Consequently, most general-aviation aircraft have flush-mounted fuselage static system ports.

Alternate Static Source

An alternate static source is a must for IFR operations. Many pilots haven't the faintest idea where the alternate source is located or if there is even one in the aircraft. Use of an alternate static source should be a mandatory checkout for all instrument pilots, and it should be tested in flight at least monthly.

Pressurized aircraft have alternate static ports located on the fuselage similar to the primary ports. In an unpressurized aircraft, however, the alternate static source is typically located inside the cabin, under the instrument panel, near the pilot's knee. While being located inside the cabin eliminates any possibility of it icing up, there are some problems. The ambient pressure inside the cabin will be less than that outside the aircraft, so there will be some instrument errors. The operating handbook should detail the difference; if not, the pilot should become familiar with the errors in a controlled training situation.

If your aircraft has no alternate static system and the primary system fails, the attitude indicator becomes a primary instrument and should be used with a power setting known to give appropriate airspeed for the desired phase of flight. All IFR pilots should develop a pitch-power-airspeed chart for each aircraft they fly.

It also is possible to break the glass of the VVI to let static air into the system. While you could break the glass of the airspeed indicator or altimeter and get the same results, they are both more important instruments, and permanent damage may occur when you break the glass. If you are fortunate enough to have copilot instruments, break one of them if they are on the same system. If they are on a different system, you might consider sliding over to the right seat and finishing your flight from there. Incidentally, all pilots should be able to land an airplane from either seat. While it isn't difficult to do, it can be tricky the first few landings, so you're best learning to do it when you aren't under pressure from some other problem. One note of caution about breaking a pitot-static instrument during static system failure. If you are in a pressurized aircraft, you will have to depressurize the cabin or the instruments will read incorrectly because of the high cabin pressure.

Pitot-Static System Preflight

The walkaround inspection should assure removal of the pitot tube cover if you use one. Also check the pitot tube, drain, and static ports for blockage, general condition, and alignment. Especially check the static ports for tape or other protection after the aircraft has been washed or painted. For flight into instrument conditions, turn on the pitot heat and feel the tube with your hand. Be brief, both to save the heater element and your skin; they heat up fairly quickly.

The old axiom "What you don't know can't hurt you" seldom holds true in aviation. Vents and openings—especially the pitot tube and static port—should be checked for foreign objects that might be lodged inside. A former flight instructor I knew used to have a pet preflight trick. He would put a toothpick in the pitot tube of his Piper Cherokee to see if his students actually got down on all fours and checked it. One day a lesson was canceled at the last minute and another student took the airplane up solo, toothpick and all; it was quite a ride for the novice. There are far too many stories about blockages caused by dirt, ice, snow, and other foreign objects to bypass a thorough look.

The cabin check should include the current local altimeter setting, which should cause the altimeter to read the airport elevation. An error in excess of 75 feet should be cause for grounding the aircraft. Both the airspeed indicator and VVI should read zero prior to engine start. During normal taxi all three instruments should remain fairly constant, aside from a little jiggling of the needles as the airplane bounces along. If there are noticeable changes in airspeed, altitude, or vertical velocity during taxi, there is a problem with the instrument.

During takeoff the airspeed indicator should "come alive" fairly early in the run. If the aircraft has two airspeed indicators, cross-check them when yours reaches the bottom of the white arc, which is nothing more than a fairly easy-to-recognize position when you glance across the instrument panel to the other one. If they are not in agreement, something is wrong. The vertical velocity should reflect a climb promptly after leaving the runway, but there always will be a slight inherent delay. The altimeter should show upward movement within a couple of feet.

Questions of accuracy regarding pitot-static instruments are best resolved on the ground, so the sooner you admit there might be a problem, the faster you can abort a takeoff or avoid flying into IFR conditions and find a place to land.

GYRO INSTRUMENTS

There are three gyroscopic flight instruments: the attitude indicator, heading indicator, and turn indicator. The attitude indicator, often referred to as an artificial horizon, is the most extensively used of the three.

Attitude Indicator

The purpose of the attitude indicator is to provide the pilot with a stable reference to the earth's horizon, depicting both pitch about the lateral axis and roll about the longitudinal axis of the aircraft. The heart of the attitude indicator is a gyroscope.

A gyroscope mounted inside the instrument case remains in a fixed position relative to the earth. Quite literally the instrument case, and the entire airplane for that matter, rotate around the gyro, causing the instrument to display true pitch and roll information regardless of the attitude of the aircraft or its turn rate. How that is possible can best be seen in Figure 2-11. A horizon line (artificial horizon) is fixed to the gyro so it always parallels the earth's surface. Older attitude indicators require the pilot to set, or "cage," the gyro after engine start. To do this, the pilot pulls the caging knob, which forces the two gimbals into vertical and horizontal positions, orienting the gyro with the horizon. Upon releasing the knob, the gimbals are free to rotate, allowing the gyro to remain parallel to the earth's surface.

There are several problems associated with the attitude indicator, primarily in the form of bank and pitch errors. They are most significant during shallow banks and after 180-degree turns. The instrument will indicate slightly less than the actual bank of the aircraft, and after rollout from a 180-degree turn, it will indicate a slight bank in the opposite direction. The instrument will also be slightly off on pitch, causing the uninformed pilot to fly the airplane into a shallow descent. It is imperative that the attitude indicator always be substantiated with a good cross-check of other instruments. Fortunately the errors are very small and a built-in erecting mechanism will correct them quickly once straight-and-level flight has been reestablished.

Attitude indicators are also susceptible to acceleration and deceleration errors. As the aircraft accelerates, the attitude indicator tends to give an erroneous, slight nose-up

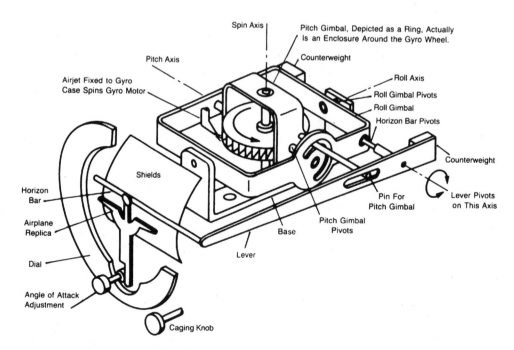

Fig. 2-11. *Air-driven attitude indicator.*

indication; deceleration has the opposite effect. It is important to realize that the attitude indicator only approximates actual aircraft attitude. However, pilots have traditionally relied excessively on the instrument, occasionally finding themselves helpless without it. Once, when flying over Atlanta on a clear morning I chanced to overhear a student pilot declare an in-flight emergency. When asked about the problem by air traffic control, the student somewhat shakily explained that his attitude indicator had just failed and he was lost over the city on his first solo flight. The controller in the tower had to explain to him that he could parallel the airplane's wing struts with the ground to make turns and he would be safe.

Heading Indicator

The purpose of the heading indicator, diagramed in Figure 2-12, is to provide the pilot with stable heading information because the free-floating magnetic compass is unreliable under any conditions other than straight-and-level, unaccelerated flight.

The heading indicator, which is not a magnetic-north-seeking instrument, must be set to a compass. The instrument needs to be reset periodically due to both random and apparent drift. Random drift is caused by bearing friction and slight imbalances in the gyro and its gimbals. Apparent drift is caused by several things. The rotation of the earth is responsible for some of the apparent drift. At the equator there is zero effect, but as the aircraft operates farther and farther away from the equator, the drift increases until at the north and south poles there is as much as 15 degrees of drift per hour.

A second cause of apparent drift is the aircraft changing positions over the earth. To minimize this, the instrument technician balances the gimbal rings to compensate for local drift. If the aircraft is flown to the opposite hemisphere, or even a different part of the country, the error may become quite pronounced. Changes in excess of 50 degrees latitude necessitate recalibration of the instrument.

It is worth noting that checking for precession (drift) is not quite as simple as one might think. The average pilot will set the heading indicator before takeoff and check it against the magnetic compass approximately every 15 minutes. Based on the comparison, the pilot makes a judgment as to the instrument's accuracy. Unfortunately, it is an inappropriate test.

To accurately check for gyroscopic precession, the pilot should turn the aircraft to the same heading used to set the heading indicator originally. This is because compass deviation varies with heading and the error you may see could be the result of a different compass deviation at the present heading. Yet another problem is compass variation— the result of crossing isogonic lines. If you are flying cross-country, this should definitely be taken into account when calculating the amount of precession.

For the pilot fortunate enough to have a synchronized gyro, there is no need to worry about precession. The gyroscopic heading indicator is electromechanically "slaved" to a magnetic sensing element. Remotely mounted, typically in the wing tip, the element is isolated from local magnetic disturbances. This provides a constant magnetic update to the gyro, preventing precession and precluding the need to reset the indicator after engine start.

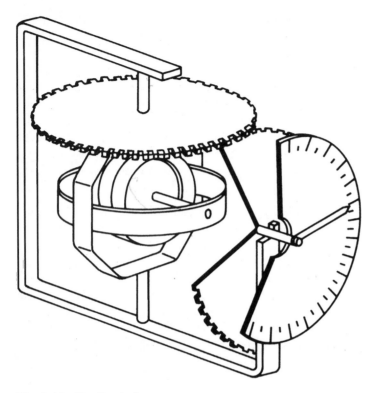

Fig. 2-12. *Heading indicator.*

Turn Indicator

The granddaddy of gyro instruments is the turn indicator. It is usually driven by a power source different from the other gyro instruments for safety-of-flight considerations. If the attitude indicator is air driven, the turn indicator will typically be electric. On some newer aircraft, one may be powered by AC while the other by DC.

The most common type of turn indicator is the turn-and-slip, or what many call the "needle/ball," as illustrated in Figure 2-13. This is actually two instruments in a single case. The turn needle senses yaw, which is rotation about the vertical axis. The turn must actually be in process before it is sensed; simply banking the airplane does not cause any indication. In most general-aviation aircraft, the instrument is calibrated for a two-minute, "standard rate turn," which is 3 degrees per second. Faster aircraft have instruments calibrated at 1.5 degrees per second, or a four-minute turn.

The ball, or slip indicator, is actually an inclinometer and not connected to the gyro system at all. Its purpose is to indicate if the amount of rudder used is correct for the angle of bank. If the ball is kept centered in a turn, it is considered a coordinated turn. The ball, usually made of black agate or steel, is placed in a curved, kerosene-filled glass tube. Since it is heavier than the liquid, the ball rolls to the low point of the tube. The liquid provides a dampening action to slightly resist the ball's rolling tendency.

In straight flight, with the airplane leaning to the right, the ball rolls to the right, indicating a slip; this is the result of one wing being low or one engine having less power. In a coordinated turn, centrifugal force and gravity combine to produce an apparent gravity, so the ball stays in the center. If there is excessive bank for the right turn, the ball rolls to the right (low) side; if there is insufficient bank, the ball rolls to the left (outside). To assure a coordinated turn, the pilot should keep the ball in the center at all times. The saying is "step on the ball," meaning that if the ball is displaced to the right, you must apply more right rudder pressure; if it is displaced to the left, you need to use more left rudder. It is also possible to decrease the bank and achieve the same results, for instance in a situation where the aircraft was already banked excessively.

A newer type of turn indicator, which is replacing the turn and slip, is the turn coordinator. This instrument senses rotation about the vertical (yaw) axis and the longitudinal

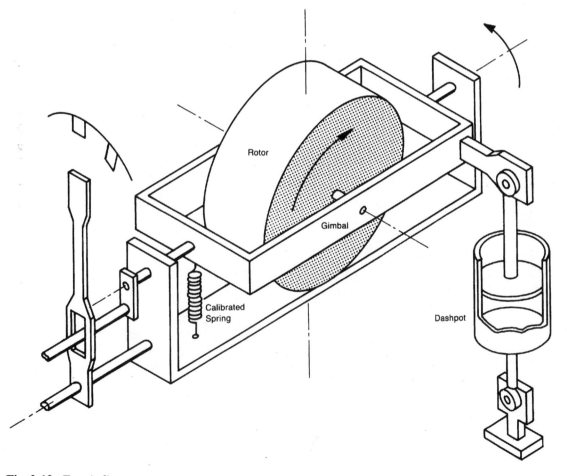

Fig. 2-13. *Turn indicator.*

(roll) axis. As soon as one wing becomes lower than the other, it senses the turn and immediately displays it. This makes it easier for the pilot to coordinate with rudder.

Rigidity in Space

All gyroscopic instruments operate under the same physical properties: rigidity in space and precession. The faster a gyro spins, the greater is its tendency to keep its spin axis in the same direction. Quite literally the gyro inside the instrument maintains a constant position while the airframe moves around it. It is this "rigidity in space" that provides a fixed point of reference for the pilot whose senses are easily misled when outside visual contact with the horizon is lost, such as on a very dark night or when flying in reduced visibility.

The gyro wheel has its bearings in a ring called a *gimbal*. Gyro instruments are gimbaled to permit the gyros to move in specific directions while preventing movement in all others, depending on the purpose of the instrument. The force that causes the gyro to spin may be either air or electricity. Older aircraft, before electrical systems, used a venturi mounted outside the airplane, as seen in Figure 2-14, as a source of vacuum. While it was Bernoulli's principle in its simplest form, there were some significant problems.

Venturi-driven systems were designed to operate at cruise, approximately 100 mph, which prevented any form of preflight check or calibration until after the aircraft was airborne. The venturi, which operates fundamentally the same as a carburetor, suffered from another formidable problem: its high susceptibility to in-flight icing similar to carb ice. There was also the problem of potential foreign object damage because the venturi was

Fig. 2-14. *Venturi.* (Photo by author, courtesy of Frasca Air Services)

exposed to the air flowing past the airframe. As aircraft became more sophisticated, the venturi was replaced with an engine-driven vacuum pump, which is still in use today. Some aircraft use a pressure system rather than vacuum, but they essentially operate the same way.

The vacuum pump draws air from the gyro case, causing a partial pressure as illustrated in Figure 2-15. The opposite side of the gyro case has an air inlet that allows cabin air, after passing through one or more filters, to enter the case through a small jet nozzle shown in Figure 2-16. The jet directs the airflow toward the gyro's rim, where there are buckets called vanes that catch it and cause the gyro to spin like a water wheel. The optimum speed varies from 8,000 to 18,000 RPM, depending on the type of gyro and who manufactured it, and is controlled by the vacuum pressure setting. Since the reliability of the instrument depends on the gyro spinning at the prescribed speed, it is important that the vacuum be set correctly. In the electric-driven gyro, the rotating mass of the gyro is typically the rotor of the motor itself forming a neat, compact package. AC-driven systems operate at speeds as high as 24,000 RPM.

Precession

What makes the gyroscope useful in flying is the fact that a rotating gyro resists any force trying to alter its spin axis, a phenomenon known in physics as "rigidity in space." However, the spin axis will move somewhat, but because of the rotation it will move 90 degrees to the applied force in the direction of rotation. This phenomenon is illustrated in Figure 2-17. In the gyroscopic heading indicator, for instance, this antiproductive force causes the heading to drift slightly. Even in straight-and-level flight, there will always be some precession due to bearing friction, but excessive precession indicates an incorrectly operating and unreliable instrument.

Gyro System Preflight

It is important to allow sufficient time for gyro instruments to spin up to the correct RPM before relying on their accuracy. For air-driven instruments, that means at least five minutes; electric driven only require about three minutes. This is especially important during cold weather operations, when lubricant and contamination within the instrument can form a sludge that may significantly resist spinning.

The attitude indicator should be monitored during taxi. It should not display more than five degrees of change in pitch or bank unless you are taxiing on very hilly terrain. There was a time when instructors taught students to make an abrupt stop while taxiing to cause the airplane to sharply pitch down. During this maneuver the pilot was supposed to watch the attitude indicator and see if the pitch changed accordingly. Such a procedure should be avoided, as any hard braking is bad for the aircraft and may cause significant bearing damage to the gyros.

The miniature airplane on the attitude indicator is adjustable vertically, allowing the pilot to calibrate it as a reference for straight-and-level flight; leave it alone when you are not flying. The previous position will probably serve as a good guide for the next flight, at least until after you level off, when you may recalibrate it. Since airspeed, load, and ambient

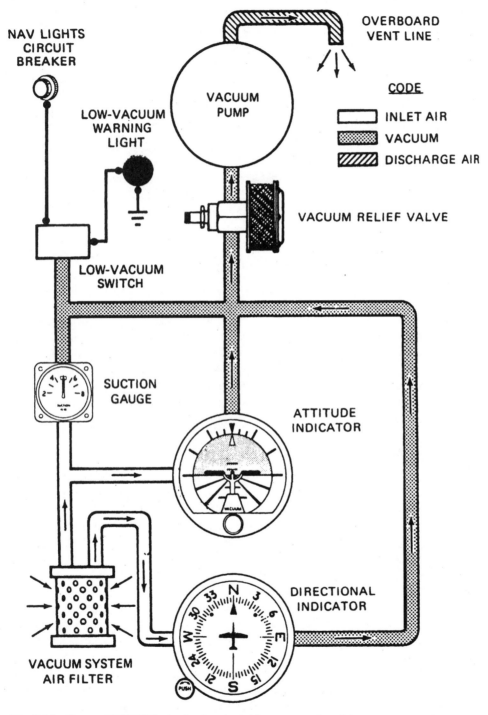

Fig. 2-15. *Cessna TU206G Turbo Stationair single-pump engine-driven vacuum system.* (Courtesy of Cessna Aircraft Company)

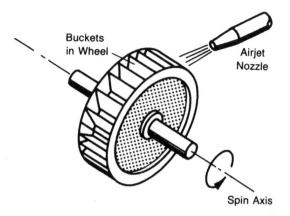

Fig. 2-16. *Jet nozzle and buckets.*

conditions will vary the exact pitch setting, it is essentially impossible to set it correctly on the ground. The procedure is to establish straight-and-level flight by reference to airspeed, and altitude then calibrates the miniature airplane. A subsequent change in airspeed will require recalibration. Many pilots will set it for straight-and-level cruise and use it as a reference for all other configurations. For instance, one bar width up may produce a 500-fpm climb at cruise power, and one bar width below may produce a 500-fpm descent. Pilots should develop a pitch-power-airspeed table for every aircraft they fly and particularly if they fly it in instrument conditions.

Prior to an instrument takeoff, never do a fast, 90-degree taxi turn. Both the heading and attitude indicators will precess as a result, and those instruments will provide unreliable information during one of the most critical phases of flight—takeoff and initial climbout. This precession lasts about one minute and can be avoided by taxiing slowly or waiting a minute while lined up with the runway before takeoff.

Preflight of the heading indicator includes setting it to a reliable magnetic reference. This should be done while on the ground, preferably prior to taxi. Once set, the pilot should observe it during taxi to assure it is changing heading appropriately. Don't taxi or fly with the heading indicator caged, as bearing damage may be the result.

Use the caging mechanism only to reset the gyro in straight-and-level flight or when stopped on the ground. If the heading immediately changes after setting, the instrument is unreliable. However, the instrument needs approximately five minutes after startup before setting the heading to assure it is up to speed.

Many pilots will align the aircraft with the runway and use the runway number to calibrate the indicator. Unfortunately, the runway number is not an accurate representation of the actual magnetic course, varying as much as five degrees. Instead, a properly calibrated, undisturbed magnetic compass should be used. Prior to taking the reading, make sure there are no pens, watches, stopwatches, or other magnetic articles lying near the compass that may introduce errors.

If you happen to have a gyro system that operates from a venturi, it is not accurate enough to set the instruments on the ground. Venturi systems are essentially unreliable

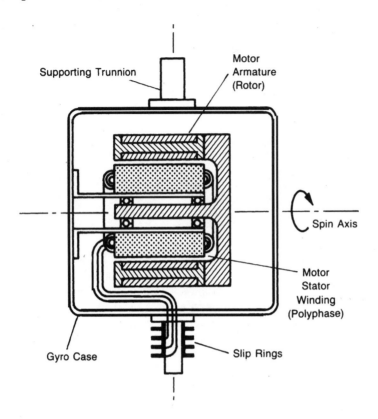

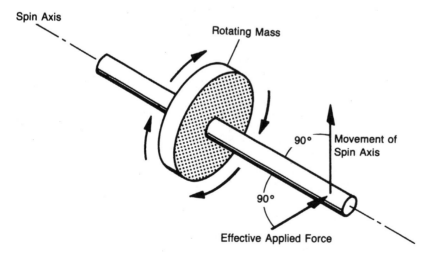

Fig. 2-17. *Gyroscopic precession.*

until you are in-flight and near cruise airspeed. This presents an additional problem to the instrument pilot because to accurately set the heading indicator, the magnetic compass must be settled in straight-and-level flight—uncommon conditions immediately after takeoff into bad weather!

All gyro heading indicators not slaved to a magnetic compass must periodically be reset to compensate for drift. Drift error in excess of three degrees per five minutes signifies impending gyro failure, and the instrument should be considered unreliable. The gyro should be reset in unaccelerated, straight-and-level flight while the compass is settled down.

You should allow the turn indicator five minutes to spin up to the correct speed before relying on it; allow three minutes for electrically powered gyros. Observe the turn indicator while taxiing. The needle should correctly indicate the direction of a turn, while the ball should move to the outside of the turn.

Preventive Maintenance

The single most important thing the aircraft owner can do to extend gyro life is to keep the air filter clean. Manufacturers tend to agree that in the long run it makes more sense to change a filter approximately every 500 hours rather than clean it. If you operate in environments where there is a lot of dust, smoke, or smog, you should increase the frequency. The filter may look clean, but dirt gets trapped inside, works its way out, and goes up the lines to the instruments. When changing the filter, you should also check the lines going out of the filter to see if they are dirty. If so, then the whole line is dirty and needs to be cleaned. The problem is that if any small particle finds its way into the gyro, it can lead to failure. Most pollution is human induced, and cigarette smoking is probably the biggest culprit, but there are others, such as face powder and cabin cleaning materials. The simple fact is that any airborne particulate matter can eat up gyro bearings.

Gyros are also extremely vulnerable to shock damage, and the part most often damaged is the bearings. Gyros should only be handled by qualified personnel, and then they should always be transported in shock-absorbent containers—even from the airplane to the shop.

For instance, if you drop a gyro from a height as little as 1 inch onto a glass table, the force is astounding. Here's why. For handling purposes, a G (gravity) is defined as the dropping distance divided by the stopping distance. Let's assume the glass flexes about .005 inches. The formula would be 1/.005 or 200 Gs! An impact of 200 Gs would certainly cause the gyro bearing races to deform when the ball bearings press against them. The long-term result would be an early failure. If the gimbal bearing takes the shock, it will lead to excessive, sporadic precession.

New gyro instruments should be kept in their original shipping container as long as possible. Use the same container for transportation to other locations or when returning it to the manufacturer for repairs. Always use the protective wrap provided, and do not remove the indicator cover anywhere other than the designated clean area of a certified repair facility to prevent instrument contamination.

When removing a gyro instrument from the airplane, wait at least 10 minutes after shutdown to assure that the gyro has stopped spinning; never remove a gyro while it is still running. Gyros are so sensitive to shock that other objects should not be allowed to bump against the gyro either. A good rule of thumb is: if you set a gyro down on a hard object and you can hear it, you have probably set the gyro down too hard. Reputable repair shops always work on gyros with a cushioned pad underneath.

In the airplane, gyro instruments, especially attitude indicators, should be "caged" prior to doing aerobatics if the instrument has a caging mechanism. The pilot should always avoid abrupt braking of the aircraft because deceleration places heavy loads on gyro bearings, yet another good reason for a carefully planned approach and minimum braking landing roll.

Troubleshooting

Heading indicators with excessive drift (more than 3 degrees in fifteen minutes), turn indicators with sluggish response, and attitude indicators that are slow to erect and/or show inappropriate deviation from level flight could be suffering from any one of several problems. Probably the most common cause is worn bearings; it's the friction that causes excessive precession.

Such symptoms may also indicate insufficient power to the gyro. In an air-driven system, this could be an inoperative pump, improperly set regulator, leaking pneumatic lines or fittings, or an obstruction in the venturi. With an electric gyro, the culprit is probably low system voltage caused by an inoperative or malfunctioning generator. Check the vacuum gauge when there is a problem with air-driven gyros or the ammeter/warning light with the electric-driven type. If those appear normal, inspect the appropriate connections to the back of the instrument case.

If you have the old venturi system, the first thing you should do is check the venturi for blockage. Worn or damaged bearings make themselves known in another way—gyro whine. If you listen carefully right after engine shutdown, you will hear the still rapidly spinning gyros whine if the bearings are going bad.

Excessively low suction gauge readings could be the result of the pressure regulator setting being too low or a leak in a line or fitting. High readings could be the result of a high pressure-regulator setting, but it can also indicate a clogged filter. A clogged filter is similar to a straw in an extra thick milk shake. If the end of the straw clogs up, the harder you draw, the lower the pressure until the straw collapses. It's the same principle, so in this case the worst thing you could do would be to lower the pressure regulator setting; even less air would get to the gyro. A good rule of thumb is: never lower the pressure regulator valve setting without first checking the filter.

Safety Precautions

The single most important safety precaution is to know the system in your aircraft. Know where appropriate circuit breakers are, have a plan for saving battery power to operate the most essentially instruments, and know which instruments are driven by electricity and which by vacuum or pressure. For instance, if a vacuum pump shaft failed during an instrument takeoff, the air-driven gyros would slowly wind down. More than one pilot has followed an inoperative attitude indicator down. Always use a cross-check to assure proper instrument operation, including VSI, altimeter, airspeed, and a gyro instrument of a different power source. Believe what the majority of the instruments are telling you. If one instrument disagrees, it is probably wrong. Every few scans should include a check of the vacuum (or pressure) gauge and ammeter. Ideal operating vacuum gauge readings are depicted in Table 2-1.

Prior to entering actual IFR conditions, observe the gyro instruments carefully to assure correct operation. Monitor the vacuum gauge and ammeter. The time to discover problems is before you are forced to compensate for them.

Tumble limits on older heading indicators are 55–60 degrees of roll and pitch, but the newer horizontal card indicators are good to 80–85 degrees. Heading indicators do have gimbaling error on some headings, during banked turns. Typically there is a 2-degree error at 20 degrees of bank, 4-degree error at 30 degrees of bank and 10 degrees error at 45 degrees of bank. Gimbal error disappears upon return to level flight.

Older attitude indicators have tumble limits between 100–110 degrees of bank and 60–70 degrees of pitch, while the newer electric 3" horizons typically have 360 degrees of roll and 85 degrees of pitch up or down.

A case of tumbling gyros is bad news; it may cause significant damage to the gyro and in instrument conditions can be disastrous for the pilot. Be prepared for possible gyro failure or accidental tumbling. Your best insurance policy is to maintain partial panel proficiency and remember that the faithful, old turn and slip indicator is your most reliable gyro: it won't drift or tumble!

MAGNETIC COMPASS

While there have been many improvements over the years, the modern aircraft compass bears a striking resemblance to its earlier counterpart; certainly the fundamentals have not changed. Most science students have studied compass theory by floating a cork in a pan of water and placing a magnetized iron sliver or needle on top of it. Because the water has no static friction, the cork turns in response to the pull of the earth's magnetic field on the north-seeking pole of the needle. Water does have sufficient friction, however, to prevent the needle from overshooting. The cork also assures that the needle floats horizontally— a potentially significant problem the closer to the north or south pole the compass gets. This simple experiment, or variations of it, guided navigators at sea for centuries.

Table 2-1

Instrument	Minimum	Desired	Maximum
Attitude indicator	3.5	4.0	5.0
Heading indicator	3.5	4.0	5.0
Turn indicator	1.8	1.9	3.2

Chapter Two

The real function of today's magnetic compass is one of redundancy. All but replaced by the directional gyro (DG), the compass has been relegated to the role of backup, occasionally being used to reset the DG to proper magnetic orientation. Unfortunately, compass knowledge has almost become obsolete too.

One of the most neglected of all aircraft instruments, it provides the heading of the fore and aft axis of the aircraft relative to magnetic north. This should not be confused with either course or track. Course is a line drawn between two points, while track is the actual movement of the aircraft with relationship to the ground.

Pilots hope their actual ground track will be the same as the course they plotted during preflight, but in reality the effect of wind usually makes that difficult. From a planning point of view, magnetic course plus or minus computed wind correction angle equals magnetic heading. But planning to use a compass and actually using it are two very different things. The effect of existing errors is so significant that a thorough understanding of them all is necessary to make the compass a reliable piece of equipment.

Earth as a Giant Magnet

The earth is a large magnet. It has poles and a flux field, but it is a bit deceiving in that the true geographic north and south poles (those at the top and bottom of the globe) are not the magnetic poles, as can been seen in Figure 2-18. Magnetic north, located in Canada, changes position ever so slightly every year. That is one reason why maps and aeronautical charts are laid out according to unshakable true north. The flux field around the earth is nothing more than a giant version of a small magnet. Flux lines come vertically (an angle of 90 degrees) out of the south pole, bend around until they run parallel to the average earth's surface at the equator, then curve back and vertically reenter the earth at the north pole. At points in between individual poles and the equator, flux emerges and enters the earth at angles of less than 90 degrees. The imaginary angle between the flux and the earth's horizontal plane is called the *dip angle*.

Imagine yourself standing thousands of feet up in the air with no magnetic objects within miles. Then suspend a long, rectangular magnet by a thread located right at the balance point. The first thing you would notice is that it aligns with the earth's flux field, pointing toward the magnetic north pole. The next thing you would probably notice is that, with a single exception, one end of the magnet is pointing downward, as if out of balance. In the northern hemisphere it would be the north-seeking end, which is incidentally the south pole of the magnet (remember, opposites attract). In the southern hemisphere, the south-seeking pole would be dipped. Only at the equator would the magnet appear to be balanced, parallel to the earth's surface.

The greater the dip angle, the more severe the problem with compass accuracy. Taken to the extreme, as when directly over the north pole, the compass needle wants to point straight down into the ground. It would obviously be difficult to navigate with the needle pointing straight down, so knowing the direction of the lines of flux alone is insufficient, you need to know the direction of the flux lines relative to a horizontal plane. That defines the direction relative to magnetic north.

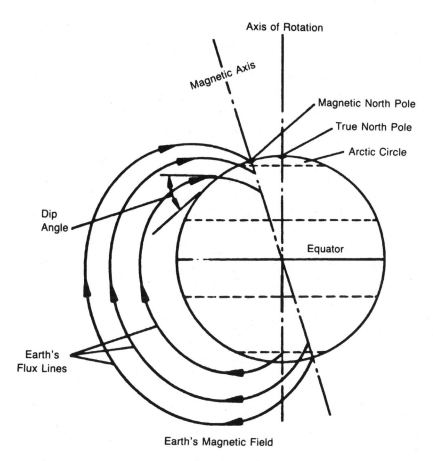

Fig. 2-18. *The earth's magnetic field.*

Compass Components

Considering the simplicity of the concept, compass design is surprisingly involved. The modern compass is a compromise of numerous conflicting requirements. Though there is some variation in manufacturing, essentially it is built as shown in Figure 2-19. The magnets, which you would expect to be big and strong, actually are not. A heavy magnet has a tendency to bob and move around excessively because of inertia. Also, a large magnet has too big of a flux field of its own, which will be greatly influenced by magnetic objects near it. By using two slender, long, parallel magnets, one on each side of the pivot, both the flux field and inertia are kept relatively small. It is the purpose of this very hard pointed pivot to mount the magnets on as friction free a point as possible so they can swing freely and align with the earth's flux field.

The pivot fits into a virtually friction-free, very hard jewel cup. Here is yet another reason for moderation in flying technique. Hard landings, fast taxi over rough surfaces, and abrupt maneuvers in flight will all cause bouncing and vibrations, the major cause of

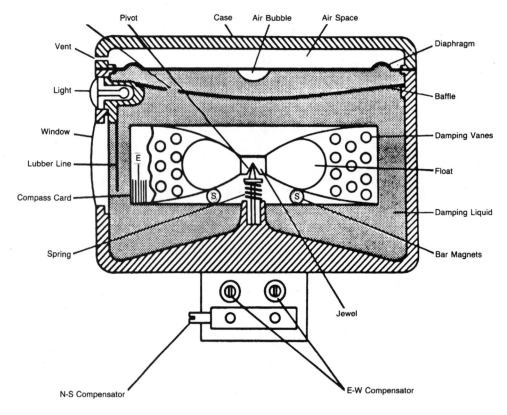

Fig. 2-19. *Aircraft compass elements.*

point blunting. As the point loses its sharpness, the compass loses its reliability. As a result, the compass card assembly is often mounted on a spring to minimize the problem, but the best preventive maintenance is still gentle handling of the aircraft. On larger compasses, the weight of the magnets and card assembly alone is sufficient to put a significant load on the pivot, so manufacturers often use a float with just enough buoyancy to relieve friction without actually floating the magnet and card assembly in the liquid.

The aviation compass card, typically a lightweight metal cylinder, has the cardinal headings depicted as N, S, E, W. The scale in between them is marked every 30 degrees with the appropriate number. The last zero of these numbers is omitted, so 30 degrees is marked by a 3, 180 by an 18. Between the numbers there are small tick marks indicating 5-degree graduations. Because the pilot reads the card from behind the compass, the markings are actually printed on the opposite side of the card. For example, when flying north, the letter W is on the right-hand side of the card, which is actually the east side of the airplane—confusing, but logical. To aid the pilot in accurately reading the compass, a vertical wire or reference mark called a *lubber line* is placed on the front of the instrument behind a glass window. It corresponds to the longitudinal axis of the aircraft, and the aircraft's magnetic heading falls directly under it.

To maintain the horizontal reference, which is necessary to assure accuracy, the entire assembly (card, magnets, and float) must hang below the pivot point. If it sits too high, the magnets will succumb to dip error, and the card will be excessively off the horizontal plane when flying in areas of high latitudes. On the other hand, if it is hung too low, it will be excessively pendulous, swinging easily during turns and acceleration. Even under the best of conditions, in the northern hemisphere the north-seeking end of the magnets will dip slightly; the opposite will be true in the southern hemisphere. To reduce the problems, the compass is filled with a liquid; this is known as compass damping.

Compass Fluid

Filling a compass with liquid sounds like a fairly simple thing. Actually, the requirements for this liquid are complex. To keep the size of the float to an absolute minimum, the liquid must have a high specific gravity. It must not get cloudy with age—very common for many liquids. Besides not being able to freeze or have a high vapor pressure, it must also maintain a relatively constant viscosity throughout the extreme temperature changes to assure continuous damping of rocking and rotational oscillation of the card.

The fluid used in a compass must be both nontoxic and flame resistant. Originally, alcohol was used, but eventually manufacturers went to both kerosene and trichlorethylene. Currently it is common to use either acid-free kerosene or silicone fluids because they not only have the dampening effect but also lubricate the pivot point. In some compasses, to prevent overswing during magnet alignment (remember fluid has very low friction), vanes are used as a sort of sea anchor. But when rolling out of a turn, momentum keeps the fluid turning in the compass, which tends to push the vanes (and the attached compass card) along with it. This phenomenon is very similar to the fluid in the middle ear, which leads to vertigo under the same situation. To reduce this problem of momentum, holes are drilled in the vanes to allow the fluid to pass through while leaving sufficient vane to damp overswing.

Some small compasses do not have liquid dampening. These are damped by magnetic eddy-current. To do this, the manufacturer uses the flux from the north-seeking end of the magnet. Because of the need for a higher than normal flux field, magnets with unusually high power-to-weight ratios are used. If doing away with the liquid sounds like a good idea, consider the consequences. The compass, which now has a greater flux field surrounding it, becomes highly susceptible to cockpit magnetic disturbances. Also, the heavier weight and lack of lubrication lead to early pivot point dulling and decreasing accuracy. The advantage of such a compass is that it is relatively inexpensive.

In the liquid-damped compass, the vane assembly is fitted carefully into a leakproof case with a clear window. Then the compass is filled with the liquid, making sure there are no air bubbles. Because liquid volume varies with temperature and virtually all liquids produce some gas over time, a flexible, perforated upper baffle is installed. The holes allow air to penetrate into an upper chamber, keeping the compass card chamber filled with liquid. In fact, the upper chamber is commonly filled with air to provide a variable pressure compensator conceptually similar to a hydraulic system accumulator nitrogen precharge. Some manufacturers put a diaphragm at the back of the compass, which works fundamentally the same. Now that you understand how the compass works, it's time to consider operational compass errors.

Compass Errors

There are two types of compass errors: static and dynamic. Static errors are the result of deviation—compass accuracy degraded by local disturbances. The source of deviation can be anything in the cockpit that will have a magnetic effect such as motors, any magnetic objects installed in the aircraft, iron or steel structure, and actual magnets used in other equipment. There is also pilot-induced deviation, such as laying a metal clipboard or a stopwatch on the instrument panel. And, of course, any electric, current-carrying wire, especially one twisted in the shape of a loop—think of all those radio leads behind the instrument panel!

To minimize deviation, a compensator is located next to the north-seeking magnets. It consists of two sets of one or more small permanent magnets that can be adjusted to compensate for local disturbances. By turning the compensator screws, these small magnets are rotated and their flux fields interact predictably with the compass's magnets. One set is for north/south adjustment and a second for east/west. These should only be set by a mechanic, as the procedure is done for a specific compass and aircraft under specified conditions. The procedure is done entirely on the ground with the aircraft configured for normal flight—radios and electrical equipment on. Called "swinging the compass," it is accomplished in the following manner:

First the aircraft is taxied onto a compass rose. Many airports have them painted somewhere on a taxiway or ramp surface. The aircraft is faced toward magnetic north according to the compass rose. The mechanic adjusts the north/south compensator until the compass reads correctly. The aircraft is turned toward magnetic east, and the east/west compensator is adjusted until the compass reads correctly. Similarly, the aircraft is turned toward magnetic south, but this time the compensator is adjusted until the compass reads halfway from its current reading to where the compass would actually read south; you are now splitting the error difference between north (which was set accurately) and south. The aircraft is again turned, this time toward magnetic west, and the difference is split between west and east. Going back to north again, the process of "splitting the difference" is repeated for north, south, east, and west one more time. The result is the best compromise possible for the deviation that exists in the cockpit. Other methods of swinging a compass without a rose include using a handheld direction-finding instrument called a *pelorus* or a true, north-seeking gyro compass.

At best, compensators remove only part of the deviation, so after the compass has been swung it is important to document the error for the pilot. This is done by filling out a compass correction card. To do so, the aircraft is turned at 10-degree increments according to the compass rose. At each increment the actual magnetic compass heading is read and the difference (deviation) is listed on the correction card. For instance, when the aircraft is positioned on the compass rose at an actual magnetic heading of 10 degrees, if the compass reads higher than 10 degrees, say 12, the difference is displayed on the compass card as STEER 12. This indicates to the pilot that to fly a magnetic heading of 10 degrees, it is necessary to steer the aircraft to a compass heading of 12 degrees.

From a practical point of view, a compass in good condition is accurate only in straight-and-level, unaccelerated flight. Any other time it is plagued by any of several dynamic errors, the most obvious of which is oscillation—the erratic movement of the compass card as a result of turbulence or rough handling. Whether induced by the environment or the pilot, when the airplane bounces and shakes, the compass will behave unreliably. When flying on instruments, for many reasons, smoothness is the all-important word. Except for those flying over the equator where the earth's magnetic field is horizontal, some errors are, practically speaking, ever-present.

Turning error, which is dominant when flying either north or south, affects the compass because in a turn the compass card banks with the aircraft. In the northern hemisphere, on a northerly heading, turning to the east, for instance, the card banks to the right. Remember that the farther north the aircraft is, the further the compass magnet dips down at its north-seeking (front) end. When the centrifugal force of the bank is added to the dip, the card swings toward the west (remember, west is on the right side of the card). As the turn proceeds toward the east, this error causes the compass heading to lag way behind the actual aircraft heading. If the pilot rolls the aircraft out of the turn when the compass reads east, the aircraft will be way past the target heading. Therefore it is necessary to anticipate this and roll out early.

On the other hand, if the aircraft is on a southerly heading, the error produces a greater heading change than actually occurs (compass leads actual heading change) and the pilot must allow for the extra heading change before rolling out. For those of you with sympathies that lie south of the Mason-Dixon line, there is a little crutch to remind you of these errors that will warm your hearts: the south leads, while the north lags.

Because of the inherent difficulties this error can produce while flying solely by reference to instruments, we use the D.G. as our primary direction instrument. However, D.G.s have been known to fail, and the compass can be your sole heading instrument. In such a case it is imperative to be proficient at timed turns using a combination of a rate-of-turn indicator and the magnetic compass. As a rule of thumb, the pilot should allow for a rollout to the desired heading proportional to the latitude where the flight is taking place. For instance, when turning from south to north at approximately 30 degrees north latitude, let's say a left turn, start rollout 30 degrees prior to north PLUS another 5 degrees to allow time to roll the wings level. This means you should begin your rollout at a compass indication of 035 degrees. When turning from north to south, let's say a right turn, you must fly past south, in this case, by 30 degrees MINUS the 5-degree rollout lead or 205 degrees. So the trick is, depending on which way you are turning, to add or subtract the latitude and allow an additional 5 degrees to stop the turn. If all this makes you think perhaps it is best to try to plan all your trips so you only fly on headings of east and west, read on.

Acceleration/deceleration error occurs on headings of east and west. By now you are well aware that the north-seeking end of the compass is already dipped down slightly when the aircraft is in the northern hemisphere. When the aircraft accelerates on an east or west heading, the aft end of the compass card tilts upward, causing the card to rotate toward north (again, remember that the north indication is actually on the south side of the card). Therefore it appears that the aircraft is turning toward the north when in fact it is simply accelerating toward the east. For the opposite reason, deceleration causes the card to tilt in the opposite direction and rotate toward the south. The way to remember this error is through the acronym ANDS: Accelerate North, Decelerate South. The only compensation is to assure you are in fact not changing your heading by closely monitoring the D.G., if operative, or the turn and attitude indicators.

Variation, though not actually a compass error, is still something to be considered. Figure 2-20 shows that magnetic and true north are not the same. Therefore, a correction must be applied to the true heading to navigate by a magnetic compass. This is because charts are laid out according to true north. One of the reasons for that is because the magnetic north pole shifts continuously. Even instrument charts are laid out according to true north, but the airways are depicted as magnetic because they emanate from the navaids, which are laid out magnetically.

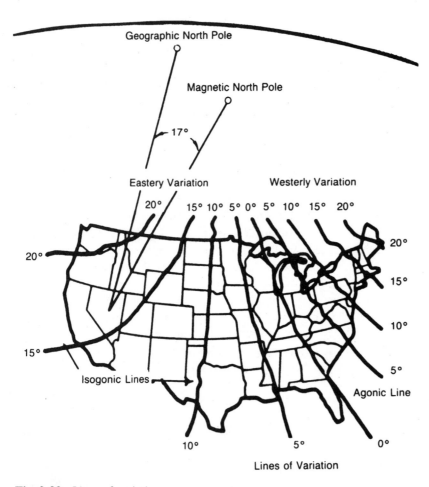

Fig. 2-20. *Lines of variation.*

The angle difference between true and magnetic depicted on charts is called *variation* and the individual lines, called isogonic lines, are labeled either E or W (for east or west). The only exception to the rule is the one line that runs directly through both true and magnetic north. That 0 line of variation is called the *agonic line*. Otherwise you add W to the true heading to get magnetic heading and subtract E from the true heading to get the magnetic one. So if you draw a line between two airports on your sectional chart and find that the true course is 310 degrees, you then look for the magnetic variation in that area. Let's say it's 10E. 310 − 10 = 300 degrees. Therefore, you must fly a magnetic course of 300 degrees. But don't forget compass deviation. Now you must consult the compass correction card. Perhaps it says for 300 degrees STEER 305. So you must actually fly a compass course of 305 degrees. All of this is assuming you are flying in the northern hemisphere; turning and acceleration errors are exactly the opposite in the southern hemisphere.

3
Aircraft Maintenance

MANY PILOTS THINK OF THE LIFE EXPECTANCY OF THEIR AIRPLANE in two ways: the airframe is forever and the powerplant goes to its TBO (time between overhaul). Inherent in that belief is the assumption that the accessories are either indestructible or, at worst, good for the life of the engine. Nothing could be further from the truth. Certainly the closer an engine gets to its TBO, the greater the scrutiny it should receive in all phases of operation. Most engines have TBOs ranging between 1,200 and 2,000 hours, but other aircraft components may have significantly shorter life spans.

TYPICAL COMPONENT LIFE EXPECTANCIES

Most propellers, for instance, are rated at less than engine TBO, commonly 1,500 hours. Fuel boost pumps range from 700–1,500 hours. Replacing the vacuum pump system filter every 100 hours, something aircraft owners rarely do, will help a dry vacuum pump to run from 400–1,000 hours; the wet type pump typically has a somewhat longer life. Mufflers frequently do not go beyond 1,000 hours, and alternators, which have a nasty habit of going out while flying in the clouds, generally give up the ghost around 1,200 hours.

Component life expectancy varies among manufacturers and is significantly influenced by the way you operate and maintain your aircraft. The trick is to have a fair idea how long these components should last and begin to cast a suspicious eye

upon them from preflight through shutdown as they approach their normally anticipated life expectancy.

REQUIRED MAINTENANCE

The first thing to understand about the airworthiness of an aircraft is that you are
responsible for it. It doesn't matter if you rented the aircraft from an FBO or borrowed
it from a friend; not owning it yourself isn't the point. FAR 91.7 Civil Aircraft Airworthiness is very clear. It states: "(a) No person may operate a civil aircraft unless it
is in an airworthy condition." and "(b) The pilot in command of a civil aircraft is responsible for determining whether that aircraft is in condition for safe flight."

This is further emphasized in FAR 91 Subpart E—Maintenance, Preventive Maintenance, and Alterations—where under FAR 91.403 General, it says: "(a) The owner or operator of an aircraft is primarily responsible for maintaining that aircraft in an airworthy
condition..."

What that exactly means to the pilot is not quite so clear, but you can rest assured that
negligence regarding the airworthiness of any aircraft you fly can result in strong action by
the FAA. On the other hand, most FAA airworthiness inspectors today have the attitude that
they are looking for a good-faith, honest effort at compliance, and if something does turn
up on an aircraft for which you're responsible, that effort carries a lot of weight with them.

So does that mean that every airplane you fly you have to check all the maintenance logs
and track airworthiness directive compliance? Few pilots actually do that, although it certainly would be the prudent thing to do. Instead, most pilots opt to find an aircraft available
to them through someone they trust. In the final analysis, though, you're the one on the hook.

The regulations have requirements for the scheduled maintenance and/or inspection
of numerous items such as altimeters, pitot-static systems, etc. The two major types of
inspections, however, are the annual inspection and the 100-hour inspection.

Annual Inspection

With any aircraft you fly, always be aware of its compliance with its annual inspection
requirement. The regulation, FAR 91.409 Inspections, is very specific. It tells us "(a) ...
no person may operate an aircraft unless, within the preceding 12 calendar months, it has
had - (1) An annual inspection in accordance with Part 43 of this chapter and has been
approved for return to service by a(n authorized) person..." Outlined in FAR Part 43 Appendix D—Scope and Detail of Items to Be Included in Annual and 100-Hour Inspections—is a very long list of required maintenance items that must be conducted
regardless of whether or not the aircraft has even flown since the previous inspection.

100-Hour Inspection

The 100-hour inspection requirement, also specified in FAR 91.409 Inspections, says:
"... no person may operate an aircraft carrying any person (other than a crew member)
for hire, and no person may give flight instruction for hire in an aircraft which that person
provides unless within the preceding 100-hours of time in service the aircraft has
received an annual or 100-hour inspection and been approved for return to service...

"The 100-hour limitation may be exceeded by not more than 10 hours while en route to reach a place where the inspection can be done. The excess time used to reach a place where the inspection can be done must be included in computing the next 100 hours of time in service." The required items for the 100-hour inspection are also outlined in FAR Part 43 Appendix D—Scope and Detail of Items to Be Included in Annual and 100-Hour Inspections.

ENGINE OVERHAUL

Looming on the horizon of every aircraft in service is engine TBO. When your mechanic tells you it's time, there are three possible maintenance solutions: Replace your engine with a new one, replace it with a rebuilt engine, or overhaul it. Replacement with a new engine can be extremely expensive and is typically not an economically viable solution. Most owners opt to have their engines rebuilt or overhauled (Figure 3-1). To make the

Fig. 3-1. *An engine awaits overhaul.* (Photo courtesy of T.W. Smith Engine Co., Inc.)

best decision for you, it is necessary to understand the terms related to engine overhauling and rebuilding.

A very commonly misunderstood term is TBO. Most pilots, being used to the concrete meaning of annual inspection and 100-hour inspection, expect to get the manufacturer's recommended TBO out of their engine. They don't understand that the emphasis is on recommended. Handling and care determine the actual time between overhauls. The pilot who practices proper engine care will almost certainly be rewarded with an actual time between overhaul in excess of the manufacturer's recommended. And yes, you can exceed TBO as long as the engine continues to operate properly and passes routine inspection! On the other hand, the negligent pilot is likely to run out the engine long before reaching TBO.

According to FAR 43.2(1), an overhauled engine is described as follows.

"Using methods, techniques, and practices acceptable to the Administrator, it (the engine) has been disassembled, cleaned, inspected, repaired as necessary, and reassembled; and (2) It has been tested in accordance with approved standards and technical data...acceptable to the Administrator."

This very vague wording requires that an overhauled engine be restored to manufacturer's approved "service limits." While there are FAA guidelines dictating service limits, it is acceptable to replace an out-of-tolerance part with a used one that is within tolerance. However, the used part may be on the borderline of the tolerance and will exceed it after only a few hours of use! From a practical standpoint, few, if any, overhaul shops are going to employ parts that are close to tolerances because they would most likely be required to replace them at their own expense during engine warranty. After the overhaul is complete, the engine logbook's total time continues, and it is noted that the engine has zero hours since major overhaul (SMOH).

While the manufacturers would prefer to sell everyone a new, or rebuilt, engine, that does not meet the needs of all customers. Therefore, most manufacturers also overhaul engines, but unlike the typical nonmanufacturer overhaul facility, they will replace all critical parts with new OEM (original equipment manufacturer) parts as a matter of routine. In addition, the OEM will use the latest modifications in accordance with manufacturer's modification changes. Some mechanics also believe that the OEM will do a better job on their own engines than an outside shop would. Lycoming, for instance, will not weld cracks in the crankcase or cylinder; instead they prefer to replace them.

What's the big deal about welding cracks in a crankcase or cylinder? According to Avco Lycoming's Williamsport Division Metallurgical Laboratory, welding aluminum can lead to significant problems. First of all, it is important to know exactly which aluminum alloy is being used, but that is proprietary information. A welder may or may not know for sure, and using an improper welding technique leads to premature failure. Yet another reason is that welding aluminum is not easy; few welders can successfully handle the job, and even the best can leave traces of tungsten, making it an unsatisfactory weld.

Complicating the problem still further, every new crankcase is heat treated for strength, then machined for exact tolerances. Normal welding technique may cause distortion of the aluminum and will certainly weaken the part. Heat treating the part again will also distort it, causing the mating surfaces to misalign. It is for that reason that welded crankcases typically leak oil. But the most skillful welder under the best circumstances still may not solve the problem. The original crack was almost certainly the result of fatigue or lack of strength in a critical area. Even after welding, porosity and invisible subsurface cracks may act as stress concentrators and cause premature failure. Dye penetrant inspection will not reveal this problem. Only X-ray produces a suitable inspection, but few operators have the equipment. All too frequently the result is failure in the same area.

Also be wary of the operator who recommends a top-overhaul; that should only be done when needed on the diagnosis of a competent mechanic. Too many engines get topped for no reason. The average engine should run to TBO if it received a good overhaul to begin with and is properly operated and well maintained.

Rebuilt Engines

The option to overhauling an engine is to purchase a rebuilt one (Figure 3-2). Rebuilt engines can only be done by the OEM because the finished engine is considered "zero time." That means it is legally a new engine, with a new serial number, new logbook, and new nameplate. In fact, it is unlikely you will get your original crankcase back. All tolerances are factory new, and only new parts are used in the process.

Choosing an Overhaul Facility

Considering the options, where to get the work done can be a difficult question. While average FBOs (fixed-base operators) don't have the equipment, expertise, or labor to perform engine overhauls, that shouldn't stop an owner from contracting with them to have the work done. Your local FBO acts as the liaison between you and the overhaul facility. You give them your airplane with the old engine and they give it back to you ready to go. If anything goes wrong later on, it's nice to be able to stroll over to your local FBO, walk up to a familiar face, and have a little nose-to-nose chat.

As liaison, the FBO will do the R&R (removal and replacement) and often overhaul the accessories while the engine is out. The actual aircraft never leaves your home airport. They handle all the engine shipping and paperwork. And because most overhaul shops give FBOs a 10%–30% discount not available to the customer, the cost is the same as if the customer took it to the overhaul shop. Finally, you are supporting your local FBO at no cost, and significantly less trouble, to you.

If you prefer to contract directly with an overhaul facility, consider doing the following. There are numerous ads in trade magazines. Contact some overhaul shops and listen to what they have to say. Then ask for, and call, a number of references. Discuss your intentions with your local mechanic. Find out where the mechanics are sending their engines for overhaul. Ask prospective companies for a copy of their warranty and look it over very carefully. A warranty says a lot about what the facility thinks of its own work.

Fig. 3-2. *Refurbished crankshaft and connecting rods ready to rebuild an engine.* (Photo courtesy of T.W. Smith Engine Co., Inc.)

A very short warranty period may indicate the routine use of used replacement parts close to their service limits. A random sampling of warranty duration ranged from 100 hours in one case to 6 months regardless of hours in another. Lycoming offers the same warranty for its overhauled engines as it does for the new ones; a strong statement indeed! After the initial warranty period expires, most facilities go to some form of prorated system where the customer pays a greater percentage as the number of hours on the engine increases.

While the warranty is important, so is proximity and immediate service. How available is the facility to solve problems after the overhaul is finished? You should be aware that most problems tend to turn up in the first 50 hours of operation. Though it is not always possible to find the right facility nearby, location is still an important consideration.

For an owner who relies heavily on aircraft availability, there is yet another option—an exchange program. The overhauling shop determines a value for your existing, run-out engine and deducts it from the price of an overhauled one, then you swap engines. Downtime can be as little as a couple of days. While the cost is less than a new or rebuilt engine, it is typically going to cost more than waiting for your own engine to be overhauled. It's worth noting that exchange programs go on the assumption that your cylinders, case, and accessories are in normal run-out condition. If, after they have had an opportunity to disassemble your engine they find unusual wear, the customer must pay the difference. Let's face it; a beater isn't worth much regardless of whether it's a car or an airplane. The same thing would happen if you were having your engine overhauled yourself. If you prefer the use of OEM parts but don't quite want to pay the price of an OEM overhaul, some operators offer two levels of service. The lowest-priced overhaul uses PMA parts, while an overhaul with OEM parts is available for a higher price.

The other major question is FAA certification. A shop does not have to be certified to do engine overhauls, although the work must be done under the supervision of an FAA-certified powerplant mechanic. Many overhaul facility owners say an FAA approval isn't necessary, that it is more of an advertising tool than anything else. One manager confided that certification was a paperwork nightmare. He said, with respect to the engine overhaul, "We don't do anything different now than we would do with certification. No matter what, the work is done under the authority of a qualified A&P mechanic." A number of FBO managers who send their own engines to various overhaul facilities echoed the sentiment. They were all more concerned about the integrity of the shop's personnel and the shop's track record than whether or not the shop was FAA certified.

As you are researching the possibilities, you will discover prices vary significantly. A good rule of thumb regarding costs would be that the least expensive option is to have an engine overhauled by an independent shop using PMA parts; using OEM parts instead will add somewhat to the same base price. Next in line would be to have it overhauled by the original manufacturer. The next most expensive would be to have it replaced with an OEM rebuilt engine, and finally to replace it with a new engine. All else being equal, you could expect the actual TBO to increase in the same order.

When preparing to make a final decision among several shops, beware of the lowest bidders; check them out very carefully. The 10-minute TBO, though not common, has happened. One unfortunate owner went to a cut-rate shop only to have them use chrome-faced piston rings with chrome cylinders—a definite no-no. The aircraft was magically transformed into the world's largest paperweight before it even got to the end of the runway as it was leaving the shop. Fortunately, 10-minute TBOs are rare, but 200–300 hour ones are not. Search for the most cost-effective shop—not necessarily the lowest bidder, but the one that does the job right for a reasonable price.

DETERMINING AIRCRAFT OR PRODUCT AIRWORTHINESS

After a newly manufactured aircraft has been certified as airworthy, many things can happen along the way to affect its airworthiness. There are two categories of guidelines for determining the airworthiness of an aircraft: type certificate data sheets and supplemental type certificates.

Type Certificate Data Sheets

Originally, an aircraft product is certified by the FAA under the specification listed in the type certificate data sheets (TCDS). If it is maintained properly, it is considered airworthy. The original specification may be modified later by an airworthiness directive (AD). Sometimes, after a new product has been on the market for a while, design flaws or operational problems are discovered. An AD is then issued requiring the owner to correct the problem as outlined.

ADs have a compliance date, after which the aircraft no longer is considered airworthy. Unless there is a very serious problem, the time allowed to make the fix is usually more than adequate. Few owners are happy to receive an AD note in the mail, but aside from some inconvenience and expense to the owner, the system works well.

Supplemental Type Certificates

The second guideline is the supplemental type certificate (STC), which is issued by the FAA when the product, or aircraft, has been altered from the original TCDS. All changes must be approved by the FAA in an STC. Once the changes are approved, the aircraft or product is considered airworthy and may fall entirely under an STC.

For instance, an STC is required if the owner of a single-engine aircraft wants to install a backup vacuum system that was not originally approved by the airframe manufacturer. Often the manufacturer of the vacuum system modification already will have the STC, which specifies the make and model of the aircraft for which it is approved.

Many pilots believe compliance with required periodic maintenance and adherence to TCDS, ADs, and STCs is all that is required to keep an aircraft airworthy, but such compliance may not be enough. Parts used for compliance also must be approved by the FAA.

AIRCRAFT PARTS

How many pilots would ground their aircraft and make an appointment with the local shop when they discover that a small screw is missing from a wing tip fairing or a bolt is missing from the seat track? More likely than not, the pilot goes to a home workshop and digs into a coffee-can collection of old screws to find a near duplicate. Such low-cost parts can be very attractive, especially when doing your own preventive maintenance as allowed by the regulations, and they would seem to present no problem. In reality, use of unapproved parts automatically invalidates the aircraft airworthiness certificate!

Fortunately, very few maintenance facilities would ever consider using unapproved parts, but the temptation to use cheap parts when possible is tremendous. There are two types of parts authorized for use in aircraft: OEM and PMA. The screw a pilot fishes out of the workshop coffee can is probably an unapproved part, but parts are not limited to such obvious origins. Some unscrupulous operators will dig into their own coffee cans and try to sell you the parts as FAA-approved. A few even will go so far as to scribe official-looking numbers on the part and/or alter it to resemble approved parts, despite potentially stiff fines resulting from such violation of the FARs. Fortunately, this is as rare as it is illegal.

More unapproved parts find their way into civil aircraft through the salvage and military surplus markets. The uninformed owner assumes that if it came off another aircraft, especially a military one, it must be acceptable. But there is no guarantee that the aircraft it was salvaged from didn't overstress or otherwise damage the part. There also is a question of storage, exposure to a harmful environment, and even the compatibility of the original manufacturing process with its intended use. There are only two types of FAA-approved parts: OEM and PMA.

OEM Parts

Original equipment manufacturer (OEM) parts are those designed and constructed for the manufacturer of the original piece of equipment, such as the elevator tab control displayed in Figure 3-3. These parts are issued a TCDS. For instance, when you purchase a brand-new engine from its manufacturer, the entire engine consists of OEM parts. Eventually, a mechanic will discover a part that needs to be replaced in that engine. When that happens, there are two legal choices that the mechanic can make: install an OEM or a PMA part. Any subsequent manufacturer who desires to build and

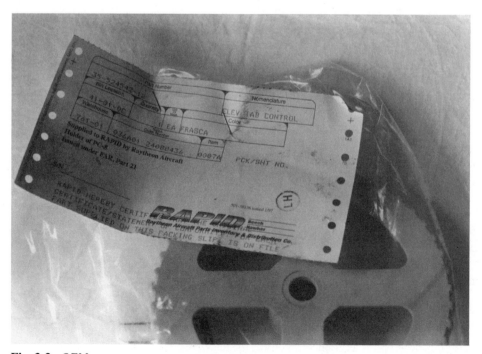

Fig. 3-3. *OEM part.* (Photo by author, courtesy of Frasca Air Services)

offer a replacement part or product must get a parts manufacturer approval from the FAA before that part is legal.

PMA Parts

According to FAR Part 45.15, PMA (parts manufacturer approval) parts must be marked with the letters FAA-PMA, name, trademark or symbol, part number, and name and model designation on which the product is eligible for installation. Parts such as a ball bearing, which are too small to be marked with an etching, must have a tag, even if it is only enclosed in a plastic bag with the part, or a similar method of identification. These methods of marking make it difficult for the unscrupulous dealer to sell unapproved parts to the aware aircraft owner.

When purchasing a part, to be certain of acceptability, you could ask for evidence of a PMA letter of approval listed by part number and eligibility for the type of product on which it may be used. Every FAA-PMA part will have one on file. Unfortunately, most distributors are unlikely to have such a letter, as it is issued to the manufacturer. Distributors might want to purchase a copy of the "Parts Manufacturer Approvals" publication from the U.S. Superintendent of Documents. Concerned aircraft owners could do the same, but the most practical course of action is simply to deal with an established and reputable dealer.

OEM Versus PMA Parts

Why should PMA parts be considered? Some manufacturers think they have come up with a better idea than the original manufacturer. They offer what they consider to be an improved part. To get a PMA, they must convince the FAA that the part is as good as the original and also must obtain an STC incorporating the changes or improvements.

There is quite a controversy between OEMs and PMA manufacturers. The problem isn't so much who makes the part. Airframe and engine manufacturers don't manufacture all of their own parts. It is not uncommon for an independent contractor to build a part under contract with the original manufacturer using the OEM blueprints and specifications. For instance, the pistons and rings that come with a new Continental engine actually are built to Continental's specifications by an independent contractor. The problem is that OEMs dislike replacement parts being built by another manufacturer as PMAs.

OEMs sometimes think PMA manufacturers typically do little or no research and testing of their own prior to getting FAA approval. Some manufacturers claim FAA enforcement of PMA quality control practices is not as strict as that for OEMs. But virtually all OEMs express concern about liability when their product is overhauled with PMA parts. It's a fact of life that when there's trouble with a component, the OEM gets sued, regardless of whose parts went in it during its last overhaul.

Some manufacturers would have you believe PMA parts are not as good as OEM, but FAA-sponsored independent research evaluating the PMA process found there was no significant problem with PMA parts. If you ask the average FAA airworthiness inspector if a PMA part is as good as an OEM, the answer is invariably, "Does the part have FAA-PMA on it? If so, it's as good as original."

But the nagging problem of liability continues to crop up and causes significant concern among OEMs. The question they rightfully ask is why should they be held accountable for the reliability of their product if parts made by someone over whom they have no control are used in the product's maintenance or repair.

The PMA Approval Process

There is a two-step process required to get PMA approval. First, there is design approval, also known as data approval. This is obtained through the FAA Engineering Aircraft Certification Office (ACO). There are three methods of accomplishing design approval: identicality, licensing agreement with the OEM, or by designing and building your own substitute part.

Identicality Method

In the identicality method, the manufacturer requesting PMA approval must prove to the FAA that the part is identical in every way to the OEMs. This is done by submitting drawings and specifications to the FAA for its review. According to many OEMs, this is where the problem comes in. They argue that these specifications are obtained through reverse engineering. The prospective manufacturer buys the original part, studies it, measures it, and develops a set of drawings and specifications. The expense of initial research and development being paid by the OEM, the new manufacturer now can sell the part for less.

Reverse engineering does take place, but when the FAA says the drawing and specifications must match the OEMs, they mean match. Any deviation at all is cause for immediate rejection, and because the specification is being compared to proprietary information the FAA cannot indicate what area or areas are unacceptable. Many things in the manufacturing process, such as hardening processes and heat treatments, are not readily discernible simply by looking at the part. Acceptance by identicality means the part will be an exact duplicate in every way.

Licensing Approval

The second method of getting design approval, a licensing agreement, is very straightforward, with drawing and specifications supplied by the OEM. This practice is common, especially for airframe manufacturers who may have a difficult time keeping up with normal production schedules, let alone spare parts.

Designing a Substitute Part

The third method of getting design approval is for the new manufacturer to design and build the substitute part from scratch. It would have to run its own tests, subjecting the part to loads, pressure, and environmental stresses to substantiate to the FAA that the new part is of equivalent strength, durability, and safety. This method also implies that the part is as good as the original.

After acceptance by one of the three methods, final approval must be made by one of the FAA's manufacturing inspection district offices (MIDO). These are not engineers; rather they are individuals who know the manufacturing processes and techniques necessary to produce the part. Their task is to evaluate the company's ability to produce the part and maintain an effective quality control program. Some PMA-seeking company officials think the process is far too demanding. One complained it took three years of processing after submitting all required data to the FAA.

Choosing the Best Parts for You

Despite all the apparent questions and concerns, there are some fairly simple guidelines the aircraft owner can follow. Above all, the single most important step you can take toward assuring optimum benefit for both your checking account and your airplane's condition is to work with an established, reputable repair facility. There is no doubt you can save some money by using PMA parts, but refer to the manufacturer's warranty first. If it says OEM parts only, then perhaps you are better off waiting until the warranty expires before using anything else. Never use unauthorized parts; it voids warranties, and when it comes to having maintenance done, shop around for the best price, but be cautious of the lowest bidder.

PREVENTIVE MAINTENANCE

Regular preventive maintenance will keep most problems in check. It is defined in Appendix A to FAR Part 43—Major Alterations, Major Repairs, and Preventive Maintenance—as "Simple preservation and the replacement of small standard parts not involving complex assemblies." It is corrective action taken before it becomes necessary to make more complex repairs.

Preventive maintenance, as outlined in FAR 43 shown in Table 3-1, may be accomplished by a certificated pilot, who is the owner or operator of an aircraft, not used in air carrier service. Most items allowed are easily accomplished by anyone with a basic mechanical aptitude, but a good rule of thumb is the following: If you aren't sure what you are doing, don't do it. Fortunately, the rule does not apply to home maintenance projects such as rewiring a microwave or even your house. You are free to destroy those things at your leisure, as it is less complicated to exit a burning house than an airborne aircraft.

UNSCHEDULED MAINTENANCE

At some point, no matter how good the preventive maintenance, no matter how much you baby your airplane, you are going to need the services of your local, friendly airframe and powerplant mechanic. If there is one thing at which a mechanic cringes, it's a pilot who says "Plane don't work right; fix it."

One hapless pilot left a note on his aircraft, saying the engine wasn't working right, and left it on the ramp in front of the shop. He returned from a business trip and found a bill of $2,000 for various engine-related work, none of which included fixing the problem he was originally concerned about.

It is important to be able to accurately define the problem and its location for the mechanic to provide quick, efficient, and reasonably priced work. The statement of the problem should reflect exactly what symptoms you observed and under what conditions

Table 3-1   FAR Appendix A—Preventive Maintenance

(C) Preventive maintenance. Work of the following type is preventive maintenance:

1. Removal, installation, and repair of landing gear tires.

2. Replacing elastic shock absorber cords on landing gear.

3. Servicing landing gear shock struts by adding oil, air, or both.

4. Servicing landing gear wheel bearings, such as cleaning and greasing.

5. Replacing defective safety wiring or cotter keys.

6. Lubrication not requiring disassembly other than removal of non-structural items such as cover plates, cowling, and fairings.

7. Making simple fabric patches not requiring rib stitching or the removal of structural parts or control surfaces. In the case of balloons, the making of small fabric repairs to envelopes (as defined in, and in accordance with, the balloon manufacturers' instructions) not requiring load tape repair or replacement.

8. Replenishing hydraulic fluid in the hydraulic reservoir.

9. Refinishing decorative coating of fuselage, balloon baskets, wing tail group surfaces (excluding balanced control surfaces), fairings, cowlings, landing gear, cabin, or cockpit interior when removal or disassembly of any primary structure or operating system is not required.

10. Applying preservative or protective material to components where no disassembly of any primary structure or operating systems is involved and where such coating is not prohibited or is not contrary to good practices.

11. Repairing upholstery and decorative furnishings of the cabin, cockpit, or balloon basket interior when the repairing does not require disassembly of any primary structure or operating system or interfere with an operating system or affect the primary structure of the aircraft.

12. Making small simple repairs to fairings, non-structural cover plates, cowlings, and small patches and reinforcements not changing the contour so as to interfere with proper airflow.

13. Replacing side windows where that work does not interfere with the structure or any operating system such as controls, electrical equipment, etc.

14. Replacing safety belts.

15. Replacing seats or seat parts with replacement parts approved for the aircraft, not involving disassembly of any primary structure or operating system.

16. Troubleshooting and repairing broken circuits in landing light wiring circuits.

17. Replacing bulbs, reflectors, and lenses of position and landing lights.

18. Replacing wheels and skis where no weight and balance computation is involved.

19. Replacing any cowling not requiring removal of the propeller or disconnection of flight controls.

20. Replacing or cleaning spark plugs and setting of spark plug gap clearance.

21. Replacing any hose connection except hydraulic connections.

22. Replacing prefabricated fuel lines.

23. Cleaning or replacing fuel and oil strainers or filter elements.

24. Replacing and servicing batteries.

Table 3-1   (Continued)

---

25. Cleaning of balloon burner pilot and main nozzles in accordance with the balloon manufacturer's instructions.

26. Replacement or adjustment of non-structural standard fasteners incidental to operations.

27. The interchange of balloon baskets and burners on envelopes when the basket or burner is designated as interchangeable in the balloon type certificate data and the baskets and burners are specifically designed for quick removal and installation.

28. The installations of anti-misfueling devices to reduce the diameter of fuel tank filler openings provided the specific device has been made a part of the aircraft type certificate data by the aircraft manufacturer, the aircraft manufacturer has provided FAA-approved instructions for installation of the specific device, and installation does not involve the disassembly of the existing tank filler opening.

29. Removing, checking, and replacing magnetic chip detectors.

30. The inspection and maintenance tasks prescribed and specifically identified as preventive maintenance in a primary category aircraft type certificate or supplemental type certificate holders approved special inspection and preventive maintenance program when accomplished on a primary category aircraft provided:

> (i) They are performed.by the holder of at least a private pilot certificate issued under part 61 who is the registered owner (including co-owners) of the affected aircraft and who hold a certificate of competency for the affected aircraft (1) issued by a school approved under Section 147.21 (e) of this chapter; (2) issued by the holder of the production certificate for that primary category aircraft that has a special training program approved under section 21.24 of this subchapter; or (3) issued by another entity that has a course approved by the Administrator; and (ii) The inspections and maintenance tasks are performed in accordance with instructions contained by the special inspection and preventive maintenance program approved as part of the aircraft's type design or supplemental type design.

31. Removing and replacing self-contained, front instrument panel-mounted navigation and communication devices that employ tray-mounted connectors that connect the unit when the unit is installed into the instrument panel, (excluding automatic flight control systems, transponders, and microwave frequency distance measuring equipment (DME)). The approved unit must be designed to be readily and repeatedly removed and replaced, and pertinent instructions must be provided. Prior to the unit's intended use, an operational check must be performed in accordance with the applicable sections of part 91 of this chapter.

32. Updating self-contained, front instrument panel-mounted Air Traffic Control (ATC) navigational software databases (excluding those of automatic flight control systems, transponders, and microwave frequency distance measuring equipment (DME)) provided no disassembly of the unit is required and pertinent instructions are provided. Prior to the unit's intended use, an operational check must be performed in accordance with applicable sections of part 91 of this chapter.

---

you observed them. Telling your mechanic that the airplane has a vibration is like telling your doctor you're sick. You must be specific. A low-frequency vibration coming from under the airplane in icing conditions points to a very different problem than a high-frequency vibration in a cruise descent coming from the area of the engine cowling.

Pertinent information should include the following:

1. A precise description of the problem.
2. The severity of the problem.
3. The power setting at the time of the problem.
4. The engine instrument indications during the problem.
5. The corrective measures you attempted and their results.
6. The flight condition at the time of the problem: cruise, slow flight, descent, etc.
7. The approximate gross weight and loading.
8. The outside air temperature.
9. The presence of visible precipitation: rain, ice, snow, etc.
10. The type and severity of any damage.
11. The known causes of damage.

Another pet peeve of many mechanics is the pilot who says, "There's a rivet missing. See ya after lunch." In describing problems, be specific about locations and use correct terminology such as forward or aft, port or starboard (left and right are acceptable, as viewed from the pilot's seat), upper or lower, inboard or outboard, and leading edge or trailing edge. For example, there is a 1-inch crack on the lower outboard trailing edge of the port aileron.

An airworthy flight results from a thorough preflight. Check for signs of security and attrition, practice preventive maintenance, be aware of airframe and engine performance during flight, and when necessary, get proper maintenance by a qualified mechanic.

4
Powerplants

THE ORIGINAL EFFORT TO DEVELOP AN INTERNAL COMBUSTION engine is documented as far back as the 1820s, but it wasn't until 1876 that August Otto and Eugen Langen of Germany built the first four-stroke-cycle engine. Through the years there have been many changes in engine shapes and the numbers of cylinders per engine. "Round engines," such as the seven-cylinder radial engine used in U.S. Navy training aircraft and shown in Figure 4-1, were very popular for many years.

There have even been 12-cylinder engines such as the V-12, shown in Figure 4-2, used by the British in one of the versions of their Spitfire. And though there have been numerous, major improvements in technology and increases in efficiency since both of those engines were in everyday service, the "Otto-cycle" engine clearly traces its way through engine history and is the undisputed great grandfather of today's aircraft and automotive engines.

The reason an aircraft engine works is based on the simple chemical fact that if you compress the correct fuel/air mixture and ignite it, the result will be a rapid, even burning, which results in substantial expansion of the gasses. To understand how this works, let's look at a typical four-cylinder engine, such as the one shown in Figure 4-3, which uses a normally aspirated carburetor. We'll follow a drop of fuel from the fuel tank through the exhaust.

Fig. 4-1. *Naval N3N with Wright 760 radial engine.* (Photo by author, courtesy of Frasca Air Museum)

Fig. 4-2. *Spitfire FR-18 with Rolls Royce Griffon engine.* (Photo by author, courtesy of Frasca Air Museum)

Fig. 4-3. *Lycoming four-cylinder horizontally opposed engine.* (Photo by author, courtesy of Frasca Air Service)

Let's call our big, old, fat drop of fuel Fred. So, Fred flows into the carburetor, where he is mixed with air in just the right proportion to support combustion. If the mixture is too rich (excessive fuel for the amount of air) or too lean (insufficient fuel for the amount of air), the engine won't run. The carburetor sets the maximum rich and maximum lean limits. Within those limits the pilot can control the airflow into the carburetor with the throttle and the fuel flow with the mixture control. In fact, with the aircraft engine, we shut it down by pulling the mixture control all the way back to cut off the fuel supply. Starved of fuel, the engine simply stops running.

As the air flows through the venturi throat of the carburetor, it creates a lower pressure, causing the fuel to be drawn in with it. If the pilot wants the engine to run faster, the throttle is advanced, causing more air to go through the carburetor, which in turns brings in more fuel.

From the carburetor, Fred and his air molecule friends travel via an intake manifold to the engine. There are four doors in the manifold called intake valves—one per cylinder. The doors open and close at separate, timed intervals. Timing is governed by when their respective cylinder is ready to receive a charge of fuel/air mixture. Fred and company happen to arrive just as the piston, which moves up and down inside cylinder #1, has begun its downward movement on the intake stroke. The intake valve opens and as the piston travels downward, it creates a partial pressure within the cylinder drawing the fuel/air mixture. As the piston reaches the bottom of its stroke, called bottom dead center (BDC), the intake valve closes and the piston begins it compression stroke.

Chapter Four

Moving upward, the piston compresses Fred and friends into a smaller and smaller space, possible because the valves are now closed, until the piston arrives at its highest point in the cylinder called top dead center (TDC). At that point, Fred and friends are now compressed into a comparatively small space with the piston at TDC; talk about losing inches around the middle. Having survived two strokes, intake and compression, the bad news is about to arrive. A spark plug delivers a spark and ignites the fuel/air mixture, an event that is simply called "ignition."

The spark plug is powered by a magneto, also "timed" in a manner similar to the valves, to provide a spark at just the right moment. Early on I said the reason an aircraft engine works is based on the simple chemical fact that if you compress the correct fuel/air mixture and ignite it, the result will be a rapid, even burning, which results in substantial expansion of the gasses. This is where that chemical reaction occurs.

As the burning gasses expand, the pressure dramatically increases, causing the piston to be pushed downward on what is known as the power stroke. This is how the engine accomplishes work. The piston itself is connected to a crankshaft by a connecting rod. As the piston is forced downward during the power stroke, its connecting rod causes the crankshaft to rotate. It is the crankshaft that is connected, often through various gears, to the propeller and a host of accessories such as the alternator and various pumps.

When the piston has reached BDC once more, the exhaust valve, again one per cylinder, is timed to open. As the piston begins its upward journey for an exhaust stroke, it forces the burned gas out the exhaust valve, down an exhaust manifold that runs along the engine in a manner very similar to the intake manifold, and finally out of the exhaust stack into the atmosphere.

So, if we look at the term four-stroke cycle, it should now make sense. A cycle is one complete series of events that occur during engine operation. During one cycle of an aircraft reciprocating engine, there are four strokes incorporating five key events. The first stroke/event is intake. The second stroke is compression with two events occurring, both compression and ignition. The third stroke (fourth event) is power, and the fourth stroke (fifth event) is exhaust.

There are one or two minor details to clarify before moving on. It is obvious where the piston gets the power for the power stroke; it is the expansion of the burning gases. What may not be so obvious is what provides power to move the piston for the intake, compression, and exhaust strokes. The answer is quite simple. There are multiple cylinders, so the power strokes of each cylinder are timed so as to occur in a sequence. For instance, as cylinder #1 is experiencing a power stroke, which turns the crankshaft, the piston of cylinder #2, which is connected to the crankshaft, may be upward on the compression stroke, while #3 may be downward on intake and #4 upward on exhaust. At all times there is one cylinder delivering power sufficient to turn the propeller and accessories and move the remaining pistons through their nonpower-producing strokes.

Another detail deals with momentum. While the above description is accurate in simple concept, the timing is not quite so straightforward. Gas flow doesn't happen instantly. Therefore, both the intake and exhaust valves actually open slightly in advance of TDC and BDC to accommodate the reality of physics.

ENGINE OPERATIONAL GUIDELINES

If the aircraft is equipped with cowl flaps, shown in Figure 4-4, they should be open during ground operations because takeoff and climb are the most difficult cooling situations for the engine. For this reason, runups should be done into the wind to maximize cooling airflow through the cowling. Always keep runups as brief as possible, using a full-rich mixture and the appropriate RPM (revolutions per minute). If you have a constant-speed prop, make sure to exercise it as outlined in the POH; this is especially important during the winter.

Don't Baby Your Engine

I have talked with far too many aircraft owners over the years who believe that if you have a long runway there's no reason to use maximum power on takeoff. They feel they are "babying" their engine and extending its life by taking off at lower power settings. Not only is that incorrect; it will actually result in a shorter TBO. Always follow the manufacturer's takeoff checklist!

In general, use full-rich mixture unless a high-density altitude exists. In that case, lean to recover power lost from an overly rich mixture. But don't baby your engine! Normally aspirated engines have been designed to use full-rated power on takeoff. Taking off with less than rated power will lead to shorter TBOs. In some aircraft, such as the Cessna 172Q, the recommendation is to maintain full power throughout the climb, while in others

Fig. 4-4. *Cowl flap.*   (Photo by author, courtesy of Frasca Air Services)

75 percent power is recommended. In either case, slightly increase climb speed on hot days by climbing at a slightly lower pitch angle—obstructions permitting—to get maximum cooling airflow through the cowling.

While the cruise phase of flight presents the lowest apparent engine workload, the pilot must be careful to avoid complacency. Seldom does an engine just fail. Usually, there are indications long before a problem arises. The vigilant pilot will use a regular scan of engine instruments and compare previous readings with current ones. Deviations should be noted and trends watched so intelligent decisions can be made early enough to allow safe, corrective action.

Generally, long descents with the engine idle should be avoided. The engine cools too rapidly, causing thermal shock and high engine wear. Start letdowns far enough out to permit carrying a little power on descent to keep engine instruments in the "green." Make the mixture richer gradually during descent until you are full rich for landing, except in high-density altitude conditions. When landing at airports with a high-density altitude, land with the mixture leaned to produce maximum power in the event of a go-around.

Proper Engine Leaning Technique

Whenever engine power is 75 percent or less, or during high-density-altitude airport operations, you should lean the engine. For taxi, lean at 1000 RPM until RPM peaks, then enrich slightly. During takeoff, use full throttle and lean until maximum RPM with a fixed-pitch prop, then enrich slightly. If you have a constant-speed prop, but carburetor-equipped engine, lean until the engine runs smooth. Fuel-injected engines should be leaned to the fuel flow setting in the POH.

The application of carburetor heat introduces warmer, less dense air into the engine, which has the effect of enriching the mixture. Use of carburetor heat also may require leaning for maximum performance. Use caution when applying carburetor heat. Some engines require it only when operating in known icing conditions, others whenever power falls outside of the tachometer's green arc. Finally, the pilot always should lean the mixture when operating in excess of 5000 feet (some POHs specify 3000 feet).

Pilots whose aircraft do not have exhaust gas temperature gauges, fuel flow meters, and other more sophisticated instrumentation must lean according to the rough engine method. This may be used with any prop as long as the engine has a float-type carburetor. First, set the throttle to the recommended power setting. Slowly lean the mixture until roughness occurs, then enrich it until the engine first runs smoothly. This will give the best economy setting.

ENGINE COOLING SYSTEMS

There are two basic cooling systems for the aircraft reciprocating engine: liquid and air. Of the two, the most efficient is liquid, but the most practical is air.

Initially, the lightest and simplest air-cooling system that was developed is what is technically termed "velocity cooling." All that means is the engine and its cylinders are sticking out into the airstream uncovered where the air rushing past can dissipate the heat from combustion. However, velocity cooling was erratic around the cylinders, especially on the aft side where little airflow reached the fins.

As engine power increased over the years, so too did the compression ratios, operating speeds, and therefore operating temperatures. Similarly, there was a demand for cooling drag reduction to help increase aircraft speed. Initially, enclosing the engine inside a cowling greatly decreased its drag, but ever-increasing demands led to smaller, tighter engine cowlings and reduced frontal areas. As engine cowlings became more aerodynamically efficient, they also diminished the volume of cooling airflow. Thus, a need for increased cooling efficiency was created, and pressure cooling was the logical solution.

Before we launch into the intricacies of pressure cooling, let's take a brief look at liquid cooling and the reason it is only rarely used in aircraft. By surrounding the cylinders with liquid, heat can be transferred by conduction to the water (or glycol, an ethylene compound used as an antifreeze), which is then pumped through a radiator. The radiator's job is to transfer the heat to the air, much in the same way as the cooling fins on air-cooled engine cylinders. As far as I know, all automobiles produced, except the original Volkswagen Beetle, use this system.

If you want to realize quickly why liquid cooling is not used in aircraft, open up the hood on your car and take a look. Note the maze of plumbing, pump, and radiator—all points of potential failure in the high-vibration environment of the aircraft. Weight is another factor; liquid-cooled engines tend to be heavier than their air-cooled counterparts. Add to the basic engine the weight of coolant (8+ pounds per gallon), hoses, radiator, and pump.

Pressure Cooling Systems

The basic principles of air cooling remain the same in the pressure-cooling system. Cylinders typically are made of chrome-molybdenum steel and cylinder heads of aluminum alloy. Very thin fins, cast or machined around the outside of both, provide increased cooling surface area for the heat to radiate out into the air. It works the same way a steam-heat radiator works in a house.

The big advantage of a pressure-cooling system is that it carefully directs the airflow within the cowling over cylinders via strategically located baffles. These baffles build up and direct the airflow so all cylinders are cooled equally from all sides. To prevent air from leaking around the baffles and taking a route less conducive to uniform cooling, rubber seals are attached to the baffles, and these press against the cowling to form an airtight enclosure.

If we imagine ourselves to be a molecule of air on a cooling journey around the engine, we must first enter the air inlet of the engine cowling. Older aircraft will probably have a high drag inlet, while newer aircraft benefit from a lower drag configuration as illustrated in Figure 4-5. In either case, the propeller is the prime mover that will push us back into the inlet when the aircraft is on the ground. In flight, ram air accomplishes the same task. From the inlet, we travel over the front-top of the engine, where very carefully positioned baffles will direct us down, around, and through the cylinder fins. In addition to assuring engine cooling, these baffles also direct other molecules to cool the oil radiator and engine-driven accessories. Now a hot little molecule, we exit through the opening in the underside of the cowling, rejoining the free airstream.

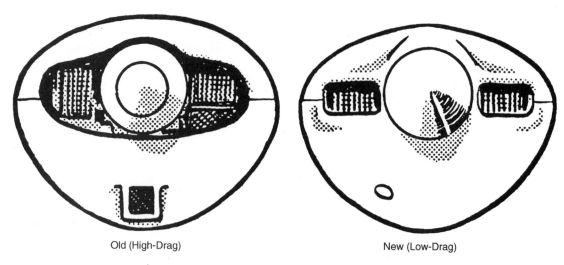

Old (High-Drag)   New (Low-Drag)

Fig. 4-5. *Air inlets of older and newer aircraft cowling.*

Engine Heat Management

As an engine's power increases, so does its heat output. Unfortunately, it often is the case that the times of greatest aircraft engine heat output also are the times least conducive to cooling airflow. Situations such as engine runup, taxi, takeoff, and climb are particular problems for larger engines because the cooling airflow is at its lowest. To permit precise control of engine temperature, such airplanes have a pilot-controlled cowl flap cooling system as depicted in Figure 4-6. This door may be adjusted from the cockpit to aerodynamically control air pressure under the cowling. The lower the pressure (cowl flap open), the greater the airflow through the engine compartment, providing maximum cooling for the engine.

As a pilot, your primary engine concern in flight is to make sure it continues to run damage-free. Excess heat is one of your engine's greatest enemies. The most critical heat-related hazard is damage to, or failure of, a piston, ring, or cylinder head. Therefore, it stands to reason you should be most concerned about the actual temperature of the cylinder and primarily the cylinder head. Unfortunately, most light, single-engine aircraft are equipped only with an oil temperature gauge. While it is true that oil, in addition to lubricating the engine, contributes significantly to cylinder cooling, oil temperature is only an indicator of cylinder temperature. Likewise, the exhaust-gas temperature gauge (EGT), with sensors installed in the exhaust stacks 4 to 6 inches downstream from the cylinders, senses engine exhaust temperature only. Too coarse to measure cylinder-temperature variations, the EGT primarily is a fuel-management device.

The primary instrument for engine heating/cooling reference is the cylinder-head temperature (CHT) gauge. The more sophisticated and expensive (CHT) systems will have a temperature probe for each cylinder, but less expensive, single-probe units still can provide the necessary basic information. If your aircraft is equipped with the more accurate bayonet-type multiprobe system, you will see temperature variations between cylinders. Cylinder head temperatures may differ by 100° F in a fuel-injected engine. Float carbureted engines can vary as much as 150° F between cylinders. This is due primarily to the fact that no cooling system is perfect. There are always misaligned baffles and variations in cowling. Also, the position of cylinders relative to the location of cowl flaps and placement of engine-driven accessories will affect the cooling of individual cylinders. The greater disparity of cylinder temperatures on engines equipped with float carburetors can be traced to the relatively inefficient fuel/air mixture distribution between cylinders.

The first law of thermodynamics essentially states that the rate of heat addition from fuel combustion equals engine power output plus the rate of heat rejection. Because their peak efficiency is only about 30 percent, aircraft engines must dissipate a lot of surplus heat.

Earlier, in tracing the route of a cooling air molecule, it was mentioned that during ground operations, the propeller pushed air back into the cowling, but in flight the prime mover was ram air pressure. A look at virtually any modern light aircraft will show that the portion of the propeller directly in front of the cooling air inlet has a very low angle of attack and generally is smaller or thicker than the rest of the propeller. It obviously is not all that effective at moving air. In addition, the actual air inlet has a fairly small opening. These two conditions significantly decrease the airplane's ability to cool the engine.

Operational Considerations

Problems resulting from improper cooling can take two forms: immediate and cumulative. Immediate problems include detonation and preignition. While significant, these problems tend to be immediately noticeable and draw attention like the proverbial "squeaky wheel."

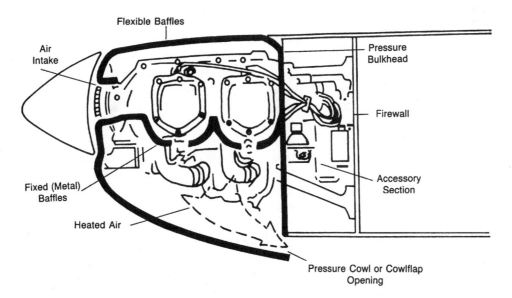

Fig. 4-6. *Engine-cooling airflow.*

Cumulative problems sneak up on the pilot and may cause permanent, sometimes catastrophic, damage before they are recognized. Here the culprit is not gross overheating of the engine once or twice, but rather the cumulative result of continuous use of improper procedures over a long period of time—each occurrence adding to the problem until there is a major malfunction. The end result is a shorter TBO by as much as half the expected life, scored pistons and cylinders, and broken or stuck piston rings.

Taxi and runups are critical periods of heat dissipation, and simple preventive measures often are ignored by pilots and mechanics. Always face the airplane into the wind when stationary, especially during engine runup. A nonmoving aircraft has a difficult time getting sufficient airflow through the cowling for cooling, so any help from the wind is beneficial. Avoid unnecessary or long runups; they generate excessive heat that is difficult to dissipate. Similarly, minimize static high-power time.

Never run up an engine under high power with the cowling removed! Frequently practiced by mechanics, this procedure leads to shortened cylinder and piston life because the cowling directs airflow to the lower and back sides of all cylinders and is primarily responsible for getting airflow to the aft cylinders.

Takeoff and climb also are critical periods for engine cooling. An engine should be operated in these phases at full-rich mixture unless the ambient conditions require leaning; the excess, unburned fuel that results from the rich mixture serves as an additional coolant. It is also imperative to long engine life that the pilot adhere to the climb speed designated in the POH. The specified speed is designed to provide the best rate of climb with consideration given to engine cooling. Lower speeds will require higher angles of attack, which translates into reduced cooling airflow and decreased TBO. A common problem associated with trainer aircraft, or any aircraft used by numerous low-experience pilots, is that the aft cylinders tend to burn out early, often significantly below TBO, because of poor climbout technique. It is very important to get a reliable mechanic to check the engine for such premature wear before purchasing a used aircraft.

Generally, cruise and descent are not considered to be cooling problems. However, one thing to watch out for is cooling the engine too rapidly during descent, particularly in larger aircraft. Shifting quickly from cruise to idle power while in a descent can lead to thermal shock, the uneven cooling of the engine, and possible engine failure. The best technique is to plan ahead and begin descending far enough away from your destination to permit a power-on descent.

Engines requiring more precise temperature control are equipped with cowl flaps. The pilot monitors the CHT and opens or closes these flaps as necessary to assure proper engine cooling. Generally, cowl flaps are open during engine start, runup, taxi, takeoff, and initial climb. They are closed after level off and remain so during cruise, enroute climbs, descents, and landing. If the ambient temperature is particularly hot, the CHT may reflect the need for partial opening of cowl flaps during enroute climb or cruise.

Pilots transitioning from less complex aircraft may be dismayed to discover that cowl flaps are yet another thing to remember. When executing a missed approach, cowl flaps need to be opened after initiating the climb because of increased power and decreased cooling airflow. When doing touch-and-goes, the pilot should open cowl flaps before applying takeoff power.

Frequently you will see pilots—sometimes encouraged by their instructors—keep the cowl flaps open while doing touch-and-goes. The reasoning goes something like this, "While it may hurt the engine to forget and leave them closed during takeoff, it doesn't hurt to leave them open and searching for cowl flaps is an unnecessary distraction when you know you are going to be doing a number of touch-and-goes." An efficiency expert might tend to agree, but an educational theorist would cringe at the thought.

The problem is that every takeoff will reinforce not using cowl flaps, increasing the probability that they will be overlooked in emergency operations. Being a pilot means learning to operate the airplane correctly under all situations, and convenient shortcuts may inconveniently lead to trouble.

Before shutting down the engine after flight, consider that you are about to end what little cooling airflow the engine is getting. Just because you stop, it doesn't mean the engine mysteriously cools off. The same laws of cooling still apply; the only difference is that there will no longer be any airflow to help dissipate the heat. The air remaining in the cowling will be heated by residual engine heat, which may require as much as several hours to cool to ambient during the summer. Meanwhile, all the fuel lines and metering devices forward of the firewall will absorb the high temperature, causing fuel expansion. As fuel expands, it is forced back to the tank, leaving vapor in the fuel lines. It is this fuel vapor in the lines and metering devices that causes problematic hot starts. Therefore, on hot days it is a good idea to avoid high power settings after landing and to idle the engine for a few minutes prior to actual shutdown to promote cooling.

ENGINE IGNITION SYSTEMS

The aircraft ignition system provides an electric spark to ignite the compressed fuel/air mixture within each engine cylinder. The FAA requires each certificated airplane with one or more reciprocating engines to have two magnetos per engine, two spark plugs per cylinder, and two wiring harnesses connecting the magnetos to the plugs.

Some magneto designs combine two magnetos within one housing. There have been serious questions about whether dual magnetos within one housing really constitute a redundant system. The two magnetos share the same housing, magneto, and common driveshaft but have independent distribution systems, coils, and points. This gives the distinct advantage of lower weight, reduced maintenance, simpler installation, and increased space availability under the cowling.

As stated previously in this chapter, aircraft engines have four strokes per cycle: intake, compression, power, and exhaust. The beginning and end of each stroke coincide with the piston positioned either at TDC or BDC within the engine cylinder. Theoretically, the spark plug should fire when the piston is at TDC at the end of the compression stroke with expanding combustion gases pushing the piston down on its power stroke. In practice, taking advantage of piston momentum, the spark plug actually fires several degrees (of crankshaft rotation) prior to TDC. This permits time for normal combustion flame propagation to increase chamber pressure along with the continued compression stroke. The end result maximizes pressure buildup slightly after TDC.

A magneto is functionally an engine-driven AC generator used to create sufficient voltage to jump the spark plug gap to assure proper fuel/air mixture combustion. Located under the engine cowling, attached to the engine itself as shown in Figure 4-7, the magneto requires no external source of electricity. Instead, it creates its own voltage through electromagnetic induction, which is simply a relative movement between a conductor and a magnetic field.

Danger of Unintentional Engine Start

A word of caution is in order regarding magnetos. A magneto presents a potentially life-threatening condition, and the pilot should always consider the magnetos to be "hot;" here's why.

It is critical to understand that what makes the magneto work, what makes it produce the necessary voltage to cause the spark plugs to fire, is simply rotating the crankshaft. The magneto is completely self-contained and doesn't require a battery or anything else before it can fire except rotation. It makes no difference to the magneto if crankshaft rotation occurs as the result of activating the engine starter or because someone turns the propeller by hand. Either one will cause the crankshaft to turn, which results in the magnetos

Fig. 4-7. *Bendix magneto mounted on an engine. (Photo by author, courtesy of Frasca Air Service)*

firing. This is one of the principle reasons why we shut down aircraft engines by starving them of fuel as we pull the mixture control into the full lean position.

If you decided to shut down the engine by turning off the mag switch, fuel would continue to be pumped into the cylinders as the engine wound down, but it would not be burned. The result would be raw fuel in the cylinders when the engine was shut down. It is true that magnetos have what is known as a "P-lead" that grounds the magneto when you turn off the mag switch, but you only need to see that frail wire one time to realize the potential for it to break.

So, thus far we have two things happening to shut down the engine. We starve it of fuel then ground the magneto to prevent ignition. Sound safe? Typically it is, but fuel systems are known to leak, especially primers, and the result could be fuel in the cylinders, even though the mixture is in the idle cutoff position. If a magneto P-lead should break, it is possible that even turning the prop a few inches could result in the engine roaring to life! As unlikely a sequence of events as that may seem, there have been quite a few cases where exactly that has happened, resulting in loss of life, limbs, and in a few cases, an aircraft taxiing out of control around a ramp with no pilot inside.

Types of Magneto Systems

There are two types of magneto ignition systems: high tension and low tension. The high-tension system, which is most common, actually generates within the magneto the high voltage necessary to fire the spark plugs. This voltage, which is transmitted through a distributor and wiring harness to the appropriate spark plug, has a tendency to jump from the harness and/or distributor and follow the path of least resistance. Known as "flashover," this is especially a problem in the low pressure and cold atmosphere of high altitudes.

Another problem with the high-tension system results from some leads being longer than others because the plugs are farther away from the magneto. These long leads tend to store energy, releasing it after the normal timed ignition. This second spark is called "capacitance after-firing." High in heat energy, it attacks spark plug electrode materials and causes approximately double the electrode erosion rate. A high-tension system is relatively lower in cost, less complicated to install than its low-tension counterpart, and lighter in weight, making it ideal for most general-aviation light aircraft.

The low-tension system overcomes flashover by transmitting low voltage from the magneto through the harness, which then is stepped up through a transformer near the spark plug. In addition to performing better at high altitudes, the magneto also has greater resistance to in-flight moisture problems, operates more efficiently, and the shorter high-tension leads reduce potential voltage loss as a result of the lead's low capacitance.

Magneto timing is critical to engine health and performance. An improperly timed spark can cause significant, even catastrophic, problems. Timing that is too advanced can lead to power loss, overheating, detonation, and preignition. If timing is too retarded, a significant power loss and increased fuel consumption would be expected. Any inspection should include checking the magneto hold-down nuts, which should be tight and safetied. If they are loose, there is a good possibility that the magneto timing will be off.

Chapter Four

The Spark Plug

The spark plug transfers high-voltage current into a cylinder and causes it to jump through the fuel/air mixture between the spark plug's center and ground electrodes, thereby igniting the mixture. If that sounds simple, then consider the conditions. The voltages are in excess of 18,000 V, there are wildly fluctuating gas temperatures as high as 3000 degrees F, and there are greatly varying pressures from partial vacuum up through 2000 psi. Come to think of it, it sounds like a drink I once had at a bar in the French Quarter of New Orleans. However, the pressure variations (in the cylinder, not the drink) are particularly significant because as compression pressure increases, the magneto voltage required to spark the gap also increases. As if these conditions were not bad enough, the plugs must not allow naturally emitted ignition radiation to interfere with sensitive navigation and communication equipment, and they must do all this reliably for long periods of time. Some time ago someone calculated that during a 100-hour operating period, any given spark plug is required to ignite approximately 7,000,000 combustion charges!

Types of Spark Plugs

Several variables are associated with choosing a spark plug. Particularly confusing is a plug's heat rating: hot and cold. The rating reflects the plug's ability to transfer combustion-chamber heat via the insulator core nose to the cylinder and engine cooling system, as shown in Figure 4-8. Hot plugs transfer heat relatively slowly, causing its insulator core to stay hotter. Cold plugs transfer heat quickly and tend to stay cooler. The correct heat rating assures the plug will operate cool enough to prevent *preignition*, which is a condition where the mixture is prematurely ignited because of some "hot spot" within the cylinder. Typical hot spots that might form in a cylinder might be a glowing hot portion of the spark plug or a bit of carbon that becomes lodged in the cylinder and glows. The correct heat rating also will assure that the plug operates warm enough to resist plug fouling by burning off unwanted contaminants.

Aircraft spark plugs use either fine-wire or massive-core electrodes. Fine-wire electrode spark plugs tend to be self-cleaning, which greatly reduces the chance for misfiring. The electrodes are actually made of precious metals, typically platinum or iridium, because these metals virtually prevent lead deposits from adhering. The plugs should be cleaned and gapped every 100 hours, and if properly maintained they can last 1700 to 1800 hours. Other benefits include easier starting, greatly reduced incidence of plug icing, and no high-altitude flashover.

A massive-core plug, as shown in Figure 4-9, should be cleaned and gapped twice as often as fine-wire electrode plugs, every 50 hours, and then the average life is in the range of 350–400 hours. It doesn't sound like much of a deal until you consider that for the typical light-duty recreational aircraft, these dependable plugs could easily last for several years and can be purchased for half the price of fine-wire plugs.

Resistor plugs reduce the heat energy of capacitance after-firing, which in turn reduces the severity of electrode erosion. Because the problem is associated primarily with high-tension magneto systems, resistor plugs will be of little use with a low-tension system. Properly used, they cut erosion in half.

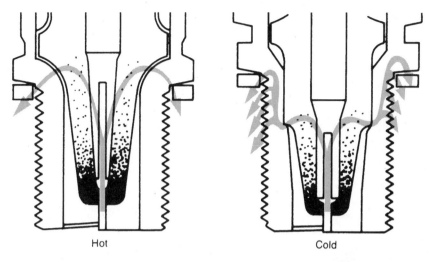

Hot   Cold

Fig. 4-8. *Spark plug heat rating.*

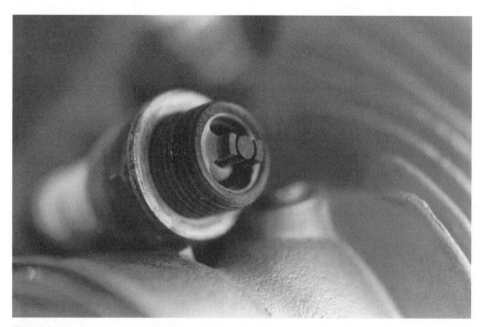

Fig. 4-9. *Massive core spark plug.* (Photo by author, courtesy of Frasca Air Services)

The reach of a spark plug is measured as the distance between the shell gasket seat to the end of the shell thread shown in Figure 4-10. Reach is determined by the cylinder-head design. The proper reach places the electrode in the best position to ignite the mixture. If the reach is too long, exposed threads easily can become "hot spots" that can cause preignition. When replacing plugs, always use a new gasket because reach includes

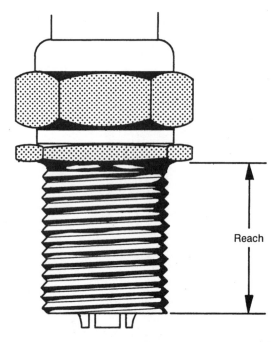

Fig. 4-10. *Spark plug reach.*

the thickness of a new gasket under proper torque. On plugs where thermocouples are used (the sensing device for some types of engine temperature gauges), no gasket is added because the thermocouple itself acts as the gasket.

Spark Plug Maintenance

Routine spark plug maintenance should be conducted every 100 hours unless otherwise stated. The plugs should be removed and checked for proper gap, fouling deposits, and general condition. Cracked porcelain usually indicates preignition. Spark plug electrodes are one of the major indicators of engine health. A good mechanic can read an electrode like an open book. Typically a brownish gray color, the electrodes readily identify problems associated with ignition, fuel mixture, and piston ring wear.

As the electrode gap widens due to erosion from normal service, there is a loss of approximately half the voltage margin between the potential available from the magneto and the voltage required by the spark plug to fire. At some point, the magneto will be unable to supply sufficient voltage to jump the ever-increasing gap. When that happens, the engine begins to misfire. Gap erosion is a normal consequence of operation, but excessive erosion can be an indication of improper fuel metering timing, magneto timing, or plug heat range.

Other causes of excessive spark plug electrode gap include capacitance after-firing and constant magneto polarity, which occur because there is an even number of cylinders. What happens is any given plug will always fire with the same polarity. Plugs that fire with a positive polarity result in excessive ground electrode wear, while negative polarity results in excessive center electrode wear. To equalize the wear of both polarity and capacitance after-firing, after removing and cleaning plugs, swap them so that top and bottom plugs change places and long and short lead plugs change places. Figure 4-11 illustrates a method of doing so.

Fouling deposits are the biggest problem in spark plug operation. Carbon fouling—which is a dry, dull black color—is the result of an excessively rich mixture, especially at idle. Spending too much time at idle and on the ground will cause significant carbon fouling of the plugs. Another possible cause is a plug with a heat range that is too cold to burn off the combustion deposits. As the carbon builds, risk of misfiring during full-power application increases. To reduce the potential for carbon deposits, operate the engine at the highest possible power setting and leanest carburetor setting conducive with safe ground-handling conditions. Periodically apply a momentary higher power to cleanse the plugs. Similarly, a faulty carburetor can lead to carbon fouling.

Oil fouling shows up as wet, black carbon deposits. Mild deposits are common on lower plugs, especially if the aircraft is seldom used, because the oil tends to drain past the rings. New engines also will experience oil fouling, indicating that the rings haven't properly seated yet; the problem should correct itself in a short time. If oil fouling appears on the upper plugs, there are several possible causes, none of which is very good. They include worn or broken rings, damaged piston, worn valve guide, sticking valve guide, and faulty ignition supply. It is especially worth noting that oil is electrically conductive, and fouling will lead to misfiring under all power conditions.

Even, fluffy fouling that ranges from tan to dark brown in color is indicative of lead fouling. It always will be present in some limited amount, but excessive lead fouling is typically caused by one of the following: high lead content in the fuel; poor fuel vaporization, which can be caused by carburetor air being too cold; operating the engine too cold; or the plug's heat range being too cold. If left unchecked, eventually lead will fill up the end of the plug and cause the plug to operate colder. This leads to misfire at high power settings, carbon collection, and then additional misfire trouble from the carbon!

Pilots seldom are aware of the delicate relationship between dust and spark plug health. Lead oxide, one of the least objectionable spark plug deposits because of its high melting point, considers silica found in dust a delicacy. It digests it and forms lead silicate with as little as 3 to 5% silica ingestion through the air induction system. Lead silicate has a melting point several hundred degrees below the original lead oxide. It causes a free-flowing spark plug contaminant, leading to misfiring at normal temperatures! The process takes from 20 to 50 hours of engine operation to develop. Therefore, it is critical to frequently clean air filters when operating in dusty environments.

During winter months, if cold air with a relatively high moisture content enters a warm cylinder during shutdown, moisture condenses on the electrodes and freezes. The ice can form a conductive bridge to the electrode and prevent the engine from starting the next time. The only options are to remove and thaw the spark plugs or apply external pre-heat to the engine.

Removal and installation of plugs is not as simple as it sounds. Aircraft spark plugs are amazingly delicate considering the conditions they are required to operate in.

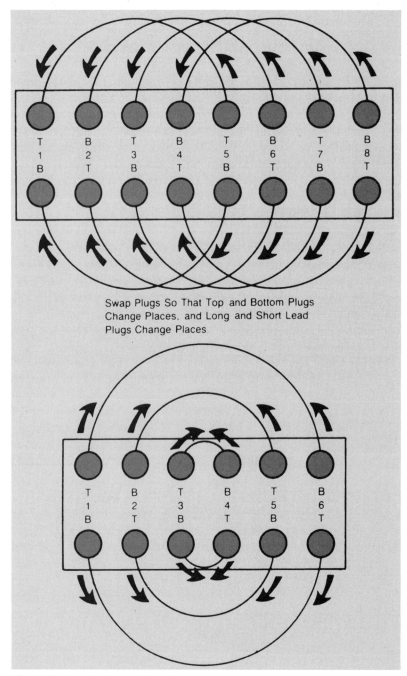

Swap Plugs So That Top and Bottom Plugs
Change Places. and Long and Short Lead
Plugs Change Places.

Fig. 4-11. *Spark plug position changes.*

Proper tools and techniques are essential. Never install a spark plug that has been dropped; always throw it away immediately. It is virtually a given that such a plug has been damaged, most probably in the porcelain, and it will be unlikely that the damage is visible.

Installation absolutely requires a torque wrench. Overtorquing will not improve the connection and may damage both the plug and the bushing. A loose plug will not transfer heat properly and may cause preignition when it overheats. Cleaning plugs requires a special tool, so you may just want to have them cleaned and gapped by a mechanic. However, Appendix A to FAR Part 43—Major Alterations, Major Repairs, and Preventive Maintenance—does allow for the owner or operator of an aircraft to replace or clean spark plugs and set spark plug gap clearance. If you opt to acquire the necessary tools to remove and service your spark plugs, here are a few tips.

When removing the plugs, put them in a tray or other device to assure you will know which plug came from the top or bottom position of which cylinder. Trays designed for the purpose are available, but an egg carton is just as effective if you label the indentations. The threads of the spark plug should also be cleaned before replacing it. Dirty threads cause poor contact between plug, gasket, and engine seat, which reduces heat transfer, causes excessive plug temperature, and increases the chance of seizing the plug.

Many mechanics apply an antiseize compound to the threads to make removal easier next time. That's fine, but never use a graphite-based compound; it acts as an electrical conductor. Instead, use only an authorized antiseize compound, antirust compound, or plain engine oil. When applying, start at least two threads away from the electrodes; otherwise it may run off and short the plug.

Conductive wiring carries electrical current between the magneto and the individual spark plugs and is protected by an insulated sheath known as *shielding*. Designed to protect the wiring from heat, atmospheric conditions, and vibration, the wiring harness also must suppress electrical interference with radio communication and navigation equipment. It is a good idea to periodically check the harness for deterioration from heat and age and the general cleanliness of terminal contact springs and moisture seals, which prevent flashover. If you are doing your own plug cleaning and gapping, don't forget to spruce up the ceramic terminal connector sleeves of the wiring harness with acetone, alcohol, or naphtha. Be very careful when handling either the ceramic or springs with your bare hand, as normal body oils left behind can act as electrical conductors and cause misfiring and flashover.

Once the engine is started, do a thorough cockpit magneto check, including the ground check. Many pilots prefer doing the ground check as the last item before engine shutdown. The plugs and harness, where visible, should be checked for general security. In addition, the harness also should be inspected for signs of aging, cracking, and chafing against the cowling. When possible, check connections for tightness.

If there was ever an aircraft system that deserved to be given top preflight priority, the ignition system is it. All the fancy equipment, retractable gear, and avionics in the world won't be very helpful if the engine stops running in flight.

ENGINE TEMPERATURE INSTRUMENTS

The fact that life may have been simpler once does not necessarily mean it was better. Aircraft engine instruments are an excellent example. Since the first tenuous flight, pilots have had some method of checking the well-being of their engines. But the crude instrumentation of bygone days often left much to be desired. Now, most modern engine instruments are both reliable and useful, provided the pilot understands exactly what is being measured and how.

As temperature increases, liquids and metals expand, though at different rates. By welding together dissimilar metals, coiling them, anchoring one end to an instrument case, and attaching the other to an indicator, you have a bimetallic (or solid) thermometer. Light aircraft commonly use this type of outside air temperature gauge. It is usually fitted right through the window or incorporated into the cabin air vent.

Another common nonelectrical temperature measuring device uses the vapor method. A gas-filled, sealed bulb and an expandable (Bourdon) tube are connected to an indicator. The bulb is located where the temperature is to be measured. The pressure inside of it varies with the temperature, causing the Bourdon tube to expand and contract, thus moving the indicator.

More sophisticated electrical temperature indicators are of two basic types: variable resistance and voltage-generated. The variable resistance temperature indicators are based on the principle that a metal's resistance to current flow varies with temperature. When a small, fixed DC voltage is applied to the sensor, some percentage of that voltage, which is determined by the amount of temperature-induced current resistance, passes through the sensor to the indicator. This type of instrument frequently is used to measure outside air temperature, cylinder-head temperature, (CHT) and oil temperature. The obvious drawback is that it requires a source of DC voltage.

The voltage-generation temperature indicator is based on the principle that certain dissimilar metals that are welded together in a loop produce a low DC voltage proportional to the temperature difference between the two ends of the loop (Figure 4-12). The thermocouple (sensor) is composed of a measuring junction where the loop is joined at the engine and reference, or cold junction, inside the instrument case. A compensating spring automatically adjusts for cabin-temperature variations, which might affect the reference end of the loop. Because metals with a very high temperature tolerance may be used, this system becomes ideal for measuring the 1500-degrees C exhaust gas temperatures of the reciprocating engine without requiring an electrical source.

Exhaust Gas Temperature Gauge

The amount of heat produced by the chemical reaction of combustion varies with the fuel/air ratio. If accurately measured, combustion heat is an important diagnostic tool for the pilot. In the early 1960s, the concept of using exhaust gas temperature (EGT) as an aid to proper mixture control became established. Prior to that the CHT was used, which is a good combustion problem indicator but lacks the accuracy and directness necessary for precise mixture leaning.

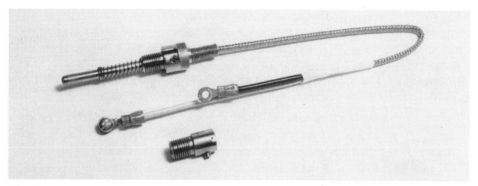

Fig. 4-12. *Cylinder head temp gauge bayonet probe.* (Courtesy of Alcor, Inc.)

In aircraft with fixed-pitch propellers, the RPM, or engine roughness, technique for leaning is used commonly. However, constant-speed propellers prevent detection of RPM variance. There is a more compelling reason for leaning with EGT, even with the fixed-pitch prop: the virtually indistinguishable difference between maximum power and maximum economy settings can be as little as .02 lb. fuel/lb. air.

A reciprocating engine can operate across a wide fuel/air mixture range—from .045 pounds of fuel per pound of air at lean misfire to .14 pounds of fuel per pound of air at rich misfire. It is at peak EGT that *stoichiometric combustion* takes place, meaning the maximum number of oxygen atoms and fuel molecules combine, producing the most efficient cruise condition. I used to love to work the word stoichiometric into conversations at parties when talking with women to impress them with how smart I was. It's probably why I've never been married. Anyway, engine operation at mixture settings leaner than that producing peak EGT can lead to cylinder and piston overheating, damage, and catastrophic failure. Operating on the rich side of peak, while less cause for concern, does produce lead-fouled plugs, costly engine deposits, and increased fuel consumption. The single-probe EGT system, common on many singles and light twins, actually measures the leanest-running cylinder as determined by the engine manufacturer.

In carburetor-equipped engines, fuel distribution differs among cylinders, and the actual cylinder experiencing the leanest mixture will vary with conditions and altitude. Because an excessively lean mixture is very harmful and the pilot has no way of knowing if the cylinder with the probe really is the leanest, a safety margin must be used to prevent inadvertent overleaning of one of the cylinders without an EGT probe. Therefore, manufacturers may recommend operating at best (maximum) power, which is achieved by enriching the mixture until the EGT is about 100 degrees F cooler than peak EGT. While safe, this technique is inefficient and costly. As the pilot leans the mixture from full rich, the airspeed will increase slightly until a temperature of 100 degrees on the rich side of peak (best power). If leaning is continued to peak EGT, the airspeed begins to decrease slightly, but range and fuel economy increase 15 percent, a significant advantage! Clearly, leaning to peak EGT is desirable but generally not feasible with the single-probe unit.

The proper technique for leaning with a single-probe EGT as displayed in Figure 4-13 is applicable to aircraft equipped with either fixed or constant-speed props. Beginning at full rich, the pilot slowly leans the mixture while watching the EGT gauge. As the mixture becomes leaner, EGT increases until the indicator peaks and then reverses. At that point, the pilot should enrich the mixture until it again peaks, stopping 100 degrees on the rich side.

The multiple-probe system, such as the one in Figure 4-14, displays an EGT for each cylinder, which permits actual determination of the leanest cylinder for existing conditions. The initially higher cost of a multiple-probe system is more than offset by fuel savings and reduced cylinder and piston maintenance. To lean, the pilot must first set cruise power for the appropriate altitude and then rotate the EGT selector knob to see individual cylinder temperatures. After determining which has the highest EGT, the pilot slowly leans to peak EGT on that cylinder. The very act of leaning will change which is the leanest cylinder, so a recheck of all cylinders is mandated. The pilot rechecks each cylinder reading and slightly enriches the mixture while monitoring the EGT gauge. If EGT drops, the cylinder was running on the rich side of peak, which is fine. The mixture is returned to its original setting. If the EGT rises, the cylinder is operating on the lean side of peak, and running leaner than the original reference cylinder. The mixture is reset to peak EGT for this cylinder and the process continued until no other cylinder EGT increases are detected. While tedious at first, practice greatly reduces the time involved, and the benefits are well worth the effort. The process must be repeated with changes in power or altitude.

The EGT gauge is divided into 25-degree F increments, with a major mark every 100 degrees and only measures relative temperature, not absolute. During preflights, check the thermocouple probes and stainless steel clamps, which are approximately 3 inches from the exhaust manifold flange of each cylinder. It also is important to check the integrity of the wire leads.

In addition to proper leaning, the EGT gauge indicates numerous in-flight problems. A decrease in both EGT and CHT indicates probable induction system blockage—perhaps icing. A decrease in peak EGT and an increase in CHT indicate detonation. A rapid increase in EGT to off scale and a rise in CHT indicate preignition. Some systems simultaneously display the EGT of each cylinder, allowing the pilot to keep a constant vigilance.

Graphic Engine Monitor

Taking things one step further is the graphic engine monitor, which simultaneously displays EGT and CHT for all cylinders. In addition, it automatically finds peak EGT for the leanest cylinder and alerts the pilot to a gradual or sudden rise in any one or all cylinder's EGT. It also is an excellent diagnostic tool for problems in ignition, fuel injection, improper fuel use, fuel distribution, and engine-failure verification in multi-engine aircraft.

Fig. 4-13. *Single-probe exhaust gas temperature gauge.* (Courtesy of Alcor, Inc.)

Fig. 4-14. *Multiprobe combustion analyzer.* (Courtesy of Alcor, Inc.)

Chapter Four

Cylinder Head Temperature Gauge

The CHT gauge, such as the twin-engine version shown in Figure 4-15, is an excellent indicator of how hot the piston, cylinder, and rings are operating. It indicates the actual temperature of what the manufacturer considers to be the engine's hottest cylinder head. A major concern for any pilot is overheating these parts, which will shorten their lives and may cause catastrophic failure. Operation at high indicated engine temperatures should be avoided because the hottest cylinder, which will vary with conditions and altitude, may not be the cylinder actually displayed. Excessively high CHT can cause detonation, engine damage, and failure. Low CHT, coupled with high power, may lead to damaged rings, pistons, and scuffed cylinder walls.

Causes for excessively high temperature include too lean a mixture, dirty fuel-injector nozzles, an induction-system leak, taxiing with cowl flaps closed, extended high-power climbs (especially at low speeds), climbs during hot ambient temperatures, idling engine with excessively flat prop pitch, and blocked cooling-air pathway or missing/broken cooling baffles.

Of the two types of CHT probes, spark plug gasket and bayonet (the element is embedded into a special well in the cylinder head), the latter is the more accurate. The CHT gauge actually is a milliammeter with a scale typically calibrated from –50 degrees to +300 degrees Celsius.

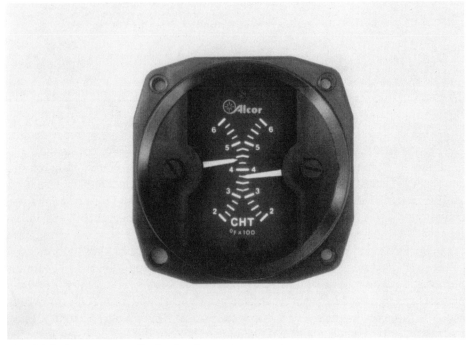

Fig. 4-15. *Twin-engine cylinder-head temperature gauge.* (Courtesy of Alcor, Inc.)

Oil Temperature

The oil temperature gauge probe is located where the oil enters the engine. Whether measured electrically or mechanically, oil temperature is displayed on an indicator divided into four ranges.

There are two red lines defining the maximum and minimum permissible oil temperatures, a green normal operating range, and a yellow cautionary range. The latter indicates a potential overheating hazard, which is an area of concern with high-viscosity oil in low-operating-temperature conditions.

Before takeoff, oil temperature should be in the green. If the temperature never rises into the green range even after a suitable warm-up period, it probably is due to a bad instrument. A takeoff may be made, provided the engine does not hesitate during full throttle application. Any hesitation should dictate an abort and further investigation.

If during climbout an excessively high temperature develops, power reduction and leveling off to increase airspeed (cooling airflow) should restore the temperature to normal range. A series of such short climbs is known as "step climbing." An excessively cold indication means insufficient lubrication for the engine and may result in dangerous power surges at high power settings.

MANIFOLD PRESSURE GAUGE

The manifold pressure (m.p.) gauge senses the absolute pressure in the engine intake manifold and displays it on a gauge, which typically is calibrated from about 10 to 30 inches of mercury (inches Hg.). Twin-engine aircraft typically will have only one instrument but with two superimposed pointers—one for each engine.

In aircraft with fixed-pitch propellers, the tachometer is sufficient to set power, but with a constant-speed prop, the RPM remains constant (within limits) while the throttle controls m.p. Normally, the engine turns the propeller, but if m.p. falls below the green arc, the windmilling prop begins to drive the engine.

Power developed is proportional to the amount of fuel burned, which is based on mass airflow to the cylinders. Airflow is difficult to measure, so intake absolute pressure (just prior to entering the intake valve) is used as the method of measurement.

For instance, if we look at the m.p. gauge when the engine is shut down, it should read the ambient air pressure, which is 29.92 in. Hg. at sea level under standard conditions. With the engine at idle, the m.p. will be very low (15 inches Hg.) because the pistons demand more mixture than the carburetor allows, creating a lower-than-atmospheric pressure. At high power, the m.p. will be 26 or 27 inches Hg. Normally aspirated engines never reach atmospheric pressure when they are running, and as altitude increases, the m.p. decreases.

When the pilot selects a higher m.p. with the throttle, the result is an increase in the fuel/air mixture entering the cylinder on each intake stroke. Supercharged engines have ambient air compressed before entering the intake manifold and are capable, especially at sea level, of producing manifold pressure several times greater than ambient. Generally, supercharged engines cannot use full throttle on takeoff or at low altitudes due to the potential for overpressuring or "overboosting" the cylinders.

According to engine manufacturer Avco Lycoming, momentary overboosts are generally acceptable, but no more than three inches Hg. for five seconds. An overboost of five inches for ten seconds indicates the engine should be inspected for possible damage, and overboosts of up to ten inches anytime mandate a teardown for detailed inspection. Anything more than ten inches requires an overhaul and crankshaft replacement. A good rule of thumb is never allow m.p. to be greater than RPM by more than a factor of four— meaning that if the m.p. is 24 inches Hg., then the RPM shouldn't be lower than 2000 (24-20 forgetting the last two zeros). Similarly, if it were 26 inches Hg., then the limit would be 2200 RPM.

Occasionally, manifold pressure gauges will behave erratically due to moisture condensation in the gauge line. To solve this problem, most manufacturers put a purge valve between the manifold pressure line and the atmosphere. Pressing the purge button opens the valve and the higher ambient air pressure enters and forces the water into the cylinder.

OIL PRESSURE GAUGE

Oil pressure, indicated in pounds per square inch (psi), is measured at the outlet of the engine-driven oil pump. Like the oil temperature gauge, the oil pressure gauge employs two red lines that indicate the maximum and minimum permissible pressure, a green arc that shows the normal operating range, and a yellow arc that indicates a cautionary range of potential hazard due to cold-start overpressure and engine-idle low pressure.

Normally, the pressure should be in the green arc within 30 seconds of engine start, slightly longer if it's very cold outside. During cold start there may be an indication of excessive oil pressure, but it should decrease when the oil warms up. If not, the engine should be shut down and the pressure valve checked for proper setting or the oil grade checked with regard to ambient temperature. Excessively high pressure leads to oil-system failure. If that occurs, engine power must be minimized until the pressure is corrected and stabilized. Prior to takeoff, the pilot always should make one last check of the oil pressure.

In flight, a fluctuating oil pressure indication probably means a malfunctioning gauge. However, it is possible the thermostatic bypass valve is not properly seated. While this is not a significant problem, a cautious eye should be kept on oil pressure and temperature to be safe, with a checkup by a qualified mechanic made as soon as practical. A low oil pressure indication in flight could be the result of several problems, the most common being insufficient oil, excessive blow-by past piston rings, and oil leaks. Other causes may be a clogged oil-pressure relief valve or incorrect setting, high oil temperature caused by improper oil grade or quantity, cooling air obstructed to oil cooler, or a dirty oil pickup screen, which restricts oil flow to the pump inlet. Because the instrument is basically reliable, if not particularly accurate, any indication of zero oil pressure should be taken seriously and warrants an immediate landing. It may be of some comfort to know that while not recommended for normal operations, most engines will develop oil pressure with as little as two to three quarts of oil. Excessive oil pressure probably is an instrument malfunction, but if it continues to remain high, a landing as soon as practical for a checkup is advised.

TACHOMETER

The tachometer, similar to its automotive counterpart, simply measures engine crankshaft RPM. In fixed-pitch propeller aircraft, it is the reference used to set engine power. Aircraft with constant-speed props use the tachometer to set the desired propeller RPM for the condition of flight. The instrument range markings are fairly standard among aircraft, with RPM limits varying somewhat. However, there are two important variations. In some aircraft, the yellow arc has a time restriction; for example, the Cessna 210N POH only permits operations in the yellow arc for five minutes at maximum power. The other is an occasional, narrow red arc located within the green arc. Operations within the red arc are not permitted except to pass through when increasing or decreasing RPM. The purpose of the red arc usually is to prevent a harmonic or resonant vibration that may lead to structural fatigue and failure. Due to the uniqueness of markings, tachometers are not interchangeable between different aircraft models.

There are three basic types of tachometers: magnetic drag, remotely driven, and electronic. The magnetic drag type is a first cousin to the automotive speedometer. One end of a flexible cable is attached to an engine-driven gear, which turns at one-half the engine RPM. The other end rotates a permanent magnet in the instrument case. Around the magnet is a drag (conductive metal) cup that is "dragged" along as the permanent magnet turns, driving a calibrated pointer. Not particularly accurate, any indication of inexplicably low power, or the inability to synchronize engines in twins, should be viewed as a reason to suspect the tachometer's accuracy if there are no other indications of engine malfunction.

Most instruments also will have a recording hour-meter designed to record time, but it is calibrated to record accurately only during cruise speed. The main source of problems is associated with the flexible double-wound drive cable. Either too much or too little lubrication with graphite will interfere with proper operation. If the indicator oscillates, the problem is likely to be either a loose casing or kinks in the cable.

The remotely driven tachometer uses a three-phase, engine-driven AC generator that drives a synchronous motor located inside the instrument. The motor, similar to the drag-cup type, turns a magnet, which operates an indicator needle. The primary advantage of this type of system is that there is no mechanical cable. Also, rather than having to rely on voltage to turn the synchronous motor, it is controlled by frequency, which is far more stable and makes the tachometer accurate through a wide RPM range.

Some aircraft have the more sophisticated electronic tachometer, which utilizes a set of special breaker points (they have no ignition function). The tach senses the opening and closing rate of the points and displays it as RPM. Multiengine aircraft commonly will have a synchroscope added to the dual tachometers. This helps trim both engines for identical power settings and aids in synchronizing engines to avoid prop "beat"—a source of both pilot and airframe metal fatigue.

FUEL QUANTITY INDICATOR

One of the most important instruments is the fuel quantity indicator, which displays fuel remaining for use in computing flying time remaining. For the light aircraft, it is very similar to its automotive counterpart—a float-type system. During preflight, it is important to compare the instrument reading visually with the actual tank level. Again, just prior to takeoff, double-check fuel quantity and selector position.

The float-type system consists of two devices: a tank (transmitter) unit and an indicator. The tank unit measures volume of fuel with a float riding on the fuel surface. An arm connects the float to a potentiometer, and as the tank empties, the arm moves across the potentiometer, which, in turn, varies the amount of voltage sent to a remote indicator. The indicator is calibrated to translate different voltages into the appropriate number of gallons remaining.

Pilots have learned to be distrustful of the fuel quantity indicator. For instance, it is sensitive to electrical system voltage fluctuations, which can cause erroneous fuel quantity readings. An even bigger problem is due to the many variations in tank design. Fuel tanks—which are integral to the wing—twist, turn, rise, and fall in an effort to take advantage of any free space where fuel might be stored. Finally, there is the problem that airplanes just don't sit still; they pitch up, down, yaw, bounce, and roll. The fuel float indicator is not unlike trying to determine sea level with a cork bobbing on the North Atlantic in winter. There are methods of reducing the problem, such as fuel tank baffles, but fluid level is not the best way to determine fuel quantity. Unfortunately, more accurate methods are very expensive and are left to larger aircraft.

FUEL FLOWMETER

As recently as 15 years ago, the fuel flowmeter was restricted primarily to the realm of large aircraft and jets. Now, many general-aviation pilots are able to glance at flowmeters, such as the one illustrated in Figure 4-16, to ascertain the rate of fuel moving from tank to engine. Flowmeters in aircraft with fuel injection actually measure the pressure across a fuel injection nozzle. From one point of view, this approach makes sense because the pressure drop across an orifice is proportional to the fuel flowing through it, and the gauge can be calibrated conveniently in gallons per hour. This method has one significant drawback: a plugged nozzle means a fuel flow decrease and a nozzle pressure increase. The gauge interprets this situation as an increase in flow, giving the pilot erroneous information that is opposite to the actual condition!

Aircraft with pressure carburetors use a hinged, spring-loaded plate called a dynamic hinged transmitter. The plate partially obstructs the fuel line, causing the flow to push against the plate as it passes; the greater the flow, the more the plate is displaced.

The plate pivots on a rotating shaft connected to the transmitter, which in turn electrically drives the cockpit indicator. The indicator usually is calibrated to show percent of horsepower and fuel flow in gallons per hour. Because there is some variability in the method employed to lean an engine using a fuel flowmeter it is important to read the aircraft POH, and follow the procedure outlined there.

Fig. 4-16. *Tru-flow 1 fuel-flow indicator.* (Photo courtesy of Alcor, Inc.)

Engine instruments have one significant similarity to flight instruments; they should be looked at as a whole rather than individually. No one instrument gives a complete picture. A potentially dangerous indication on one can be evaluated intelligently only after checking the others.

PREFLIGHT

When conducting a preflight inspection, there are several engine considerations. The cooling air pathway should be checked carefully, including air inlet, air pathway under cowling, and the air exit area or, in some aircraft, the cowl flaps.

Oil quantity, of course, is an extremely important item on the preflight check. Pilots should check their aircraft POH to determine recommended oil levels, which might differ for cross-country and local flying. Don't forget that oil grade requirements can differ with the season.

In addition to using the proper oil in cold weather, it is a good idea to pull the prop through by hand at least six revolutions to loosen up the oil. When temperatures drop to less than 20 degrees F and the airplane is not kept in a heated hangar, preheating the engine may be necessary. Normal oil pressure should come up within 60 seconds after start. If pressure does come into the green arc but fluctuates, the oil is still too cold and causes cavitation of the oil pump. Shut down the engine and preheat before another start attempt.

While under the cowling, check general engine security. A visual inspection will turn up most oil and fuel leaks, while a tug at the engine mounts will attest to their integrity. If the airplane is a rental, take a look at the valve covers. Often the most recent compression check will be written in chalk there and is a good indication of the general health of the engine (more on compression checks later).

Find a good, reliable way to determine fuel quantity, such as a fuel dipstick. Fuel gauges, notoriously inaccurate, should not be trusted. Assuring the proper type of fuel also is important. Using fuel with a higher octane rating than necessary on a regular basis may cause excessive lead fouling; too low a rating could result in detonation. The key here is the fuel color, and you can check it by draining some fuel into a plastic see-through cup. While checking the color, also look at the quality of the fuel. It should be clean, evenly colored, and transparent, without sediment, dirt, or water. Any presence of water is a problem; it is distinctly visible and will collect at the bottom of the plastic cup. Keep draining and taking samples until there is no longer a trace.

Cooling System

For the most part, preflighting the cooling system is a matter of looking for the obvious. Baffles should be periodically checked for proper alignment. Missing or broken baffles should be replaced. Cylinder and cylinder head cooling fins create the majority of engine cooling, so any problems should be brought to the attention of your mechanic.

Airflow blockage is another common problem that should be considered on every preflight. There are three areas of concern: intake, inside of the cowling, and exhaust. Probably the most common blockage is a bird's nest in the air intake or exhaust area. This can be solved by purchasing and using air intake plugs that completely fill up the air intake area, preventing birds or anything else from getting in.

The exhaust area can be conveniently closed up if the airplane has cowl flaps. Otherwise, make sure to carefully check this area because birds are just as liable to build a nest there. Less obvious is a blockage actually within the cowling. I once had a collection I would show students that included red, green, blue, and dirt-gray shop towels, an instruction manual for an engine compression tester, the beginning of a bird's nest, and a torque wrench.

Magneto System

Checking the condition of magnetos can be accomplished in two ways: grounding check and differential check. As explained previously, magnetos are connected by a primary wire (P-lead) through the magneto switch to ground. Its purpose is to provide an internal path of least resistance, rendering the magneto inoperative when the switch is turned off. If the P-lead fails, the magneto stays "hot." The significant drawback of this system is on the ground, where a "hot" mag patiently waits for someone to come along and even slightly turn the propeller, causing the impulse coupling to produce a spark that could make the engine roar to life.

To check for proper grounding of the magneto, set the throttle at idle and rapidly move the mag switch from BOTH to OFF, then back to BOTH. You should hear a momentary engine failure. If there isn't one, then one or both mags are hot and you need to find a mechanic.

Incidentally, P-leads go to specific mags and are not interchangeable. During start-up, there is insufficient piston momentum to overcome normal "advanced" sparking prior to TDC, so the starting spark is "retarded" at or near TDC on the compression stroke. An impulse coupler is used to snap the magneto through its firing position rapidly so it will produce a high-energy spark, even though the engine is being cranked slowly and simultaneously to retard the spark. In such a system, it typically is the left magneto that has the impulse coupler, while the right mag is grounded automatically during the start procedure to prevent its advanced spark from prematurely igniting the fuel/air mixture and causing the piston to kick back.

The entire system (impulse coupler, retarded left spark, grounded right magneto, and starter motor) is controlled by the starter position on the ignition switch. Switching magneto P-lead will deactivate the impulse coupler and cause the wrong mag to fire in advance TDC. The probable result will be an engine that will try to run backwards with potentially significant damage. When hand propping, never put the switch to BOTH for the same reason. Always set it to the magneto that has the impulse coupler. Even the more expensive systems that use a starter-vibrator instead of an impulse coupler operate functionally the same and require proper mag selection when hand propping.

The differential magneto check, familiar to all pilots, compares one mag to the other. Using the manufacturer's recommended power setting, ground one magneto and note the RPM difference from BOTH. Then ground the other, note its difference from BOTH, plus note the difference from the previous reading. The RPM drop is the result of incomplete combustion. Using only one set of plugs is not as efficient as using both sets, so there is a loss of engine power and RPM. The manufacturer publishes a maximum-allowable RPM drop from BOTH and a maximum allowable RPM difference between magnetos. If the RPM drops in excess of these figures, ground the aircraft.

Several other possible conditions may be noted during the differential mag check. For example:

- Immediate, excessive RPM drop probably indicates failure of the secondary mag coil, which prevents a sufficient current flow to get the spark to jump across the spark-plug gap.

- Slow, excessive RPM drop indicates a problem with magneto timing. If the engine slowly dies, it again is probably because of a failure of the secondary coil, which prevents sufficient high voltage at the lower RPM used for the check.

- No observable drop in RPM indicates probable P-lead failure.

- An engine failure during the mag check can be traced to probable failure of either the primary coil or the breaker points. Caution: do not switch back to BOTH. Unburned fuel/air mixture can build up in the cylinders and exhaust. Switching to both will ignite the mixture and possibly cause major damage to the cylinder and exhaust. Throttle to idle, then switch ignition to BOTH to prevent damage.

- A condenser failure typically shows up as high RPM miss and is not noticeable during low RPM testing. Points will prove to be pitted and burned upon inspection.

It is worth noting that operating the engine on one mag for as little as 30 seconds can lead to fouling of the inoperative plugs.

Moisture, which finds its way into a magneto simply by condensing when the temperature changes, can leave corrosive acids behind. These acids act as carbon trails begging electricity to flashover. It is a good idea to reduce the problem as much as possible by checking for cracks, loose cover plates, or other preventable methods of moisture entry.

TROUBLESHOOTING

On a hot day, when you attempt to start a fuel-injected engine that recently has been flown, it is common for it to come to life only to die swiftly. When the aircraft is parked after flight, residual engine heat "cooks" the fuel lines and metering devices under the cowling like thawing dinner by leaving it on the stovetop directly above the pilot light. This causes the fuel to expand, which forces it back into the fuel tank, leaving only vapor in the lines. Because a little fuel remains in the lines, the engine roars briefly to life, but the fuel pump is incapable of moving enough vapor to keep it running, so the engine quickly dies.

To avoid this problem, before attempting the start, you should purge the fuel lines by placing the mixture control at cutoff, putting the throttle at full open (some throttle linkages prevent high pump action with throttle retarded), and turning on the auxiliary fuel pump to high pressure for about 20 seconds. This procedure pumps cool, fresh fuel through the lines, purging the vapor and cooling the system. The fuel return system routes first the vapor, then the fuel back to the tank, leaving the lines filled with fresh fuel. To start the engine, turn off the fuel pump, place the mixture at full rich, open the throttle partially, and engage the starter; the engine should start easily.

Rough engine idle could be due to either an excessively lean/rich mixture, which would respond to adjustment, or to a mechanical problem, which should be referred to a mechanic. Possible culprits are an induction air leak, improper fuel pressure, bad compression on one or more cylinders, fouled plugs, ignition system problems, or plugged injector nozzle(s).

If the tachometer indicates an excessively low ground-idle RPM, check the carburetor heat or the prop control setting. Other possible problems are a restricted air inlet or governor out of adjustment.

If your engine consumes too much oil, one problem easily could be use of the wrong type of oil. That is tough to figure out once it is added, so never allow anyone else to add oil to your airplane. Other problems that may cause high oil consumption include worn valve guides and bad or improperly seated rings. The obvious problem in the case of a low oil pressure indication is insufficient oil quantity. More difficult problems might include failure of the pressure-relief valve or a clogged oil pump inlet.

When an engine will not develop rated power, improper use of carburetor heat and mixture top the list of likely causes. Too low a fuel grade runs a close third. If all three check out, there are several other possibilities, such as insufficient air induction or fuel flow, low cylinder compression, or incomplete ignition.

Detonation

Engines are most susceptible to detonation at high power settings, particularly if combined with improper leaning. Excessive temperature can cause the fuel/air mixture within the engine cylinders to detonate explosively. This causes a sharp, excessive pressure rise accompanied by a distinct metallic knock which, unlike automobiles, is seldom heard in an airplane. In addition, there also is a significant temperature increase in the combustion gases. This temperature rise causes the fuel/air mixture to expand, less fuel to burn, and engine power to decrease. Another reason for the power loss is the piston's inability to accelerate rapidly enough to convert the unusually high-pressure spike into power.

Perhaps the most serious aspect of detonation is its insidiousness. Cracked pistons, burned valves, and catastrophic engine failures show up only in the most severe cases. Light to medium detonation may not be noticeable at all in the cockpit but will still lead to piston, ring, and cylinder-head damage over time.

Causes of detonation are typically too low an octane fuel, excessively hot CHT, hot fuel-air mixture, excessively lean mixture, and high intake manifold pressure. Provided you have the correct fuel, detonation usually can be stopped by enriching the mixture, making shallower climbs to increase cooling airflow, selecting full-open cowl flaps, and reducing power. If none of these measures solve the problem, you should terminate the flight and seek the help of a mechanic.

Preignition

Often confused with detonation, preignition results when the fuel/air mixture is ignited prior to spark plug discharge. The symptoms are similar to (though not as severe as) detonation and include engine roughness, backfiring, high CHT, and a loss of power. Preignition may be caused by a hot spot in the cylinder, often a result of a carbon buildup on the cylinder head or spark plugs. It is also possible for a hot spot to develop if valve edges are ground too fine. In that case, the thinness causes the valve edge to glow, which, in turn, ignites the fuel/air mixture prematurely. As with detonation, significant damage may result, such as cracked pistons or valves, so it is important to reduce CHT as quickly as possible. This is done by enriching the mixture, reducing power, and maximizing cooling airflow with higher climb airspeeds and wide-open cowl flaps.

Engine Cooling

Abnormally high cylinder head temperatures in flight may be the result of a number of problems. First on the list of culprits would be an excessively lean mixture, which would be corrected by enriching the mixture. This tends to happen at high power settings during takeoff and initial climb. Similarly, any operations at higher-than-recommended power settings should be avoided. Operations during excessively high ambient temperatures also may lead to unusually high CHT. Some aircraft POHs list a maximum ambient temperature, but typically this is higher than the average pilot will ever encounter. Missing or misaligned engine baffles, air leaking out of the cowling, and airflow blockage all may cause high operating temperatures. Even a faulty ignition system can show up on the CHT gauge as too hot.

The best plan of attack for unusually high temperatures is to enrich the mixture, open cowl flaps, reduce power, and reduce drag as much as possible to keep up the cooling airflow. If none of those have the desired effect and the engine is above the safe operating temperature outlined in the POH, land the plane as soon as possible. Meanwhile, keep a sharp eye on the oil temperature gauge; if it too begins to rise beyond the maximum limit, you have a real problem on your hands and you should find a place to land immediately.

PREVENTIVE MAINTENANCE

Preventive maintenance starts with a proper preflight inspection and encompasses all aspects of engine operation, both on the ground and in the air. The best preventive maintenance is to use the engine in normal operations on a regular basis. Gaskets, seals, and O-rings need frequent lubrication to stay in condition, as long periods of inactivity lead to oil leaks. With time, oil thins and evaporates, allowing the moisture in the air to coat the cylinders and begin the rusting process. Changing the oil as recommended and flying the airplane at least once a week should take care of the worst situations.

If average humidity is below 70 percent, you need only fly the airplane once every two weeks. To be beneficial it is necessary to actually fly the airplane at cruise for at least 30 minutes each time. Anything less provides insufficient time for the oil-entrained moisture to dry out. Merely doing an occasional runup does more harm than good. It causes a dramatic temperature change in the engine that, when shut down, causes water to condense inside the cylinders, leading to rust and corrosion. Because ground operations in general aren't very good for an airplane, it shouldn't be difficult to imagine that they are pure poison as an airplane's sole weekly exercise.

While the FAA only requires private aircraft to undergo an annual inspection, next to regular use, the 100-hour inspection is the best preventive maintenance. Engines don't just quit; they gradually get sick. You'll see symptoms long before anything really nasty happens, and the 100-hour inspection looks for such telltale signs.

One of the prime indicators of improper operation and engine health is spark plugs. Every 100 hours, the plugs should be removed and checked; normal plugs will have a sort of brownish-gray color. There are three basic problems that this check will turn up: fuel fouling, lead fouling, and oil fouling. Fuel fouling is indicated by sooty, black deposits, commonly a result of not leaning during high-altitude operation (excessively rich mixture). Other conditions that may lead to fuel fouling are too rich an idle mixture, excessive ground operation, frequent power-off descents at full rich, and too low an operating temperature.

Lead fouling, indicated by gray deposits on the plugs, is normal in small amounts. Large buildups will require frequent plug cleaning and generally are the result of too high an octane fuel. Black, wet deposits, particularly on the bottom plugs, indicate oil fouling. Accompanied by high oil consumption in a high-time engine, this easily could mean that an engine overhaul is due. You probably will find excessive cylinder-wall wear, worn valve guides, or even a broken piston ring.

The air filter also should be changed at least every 100 hours, more often in high dust or smoke environments. Unlike its automotive counterpart, the aircraft filter doesn't carry dirt well and quickly leads to loss of power, excessively rich mixtures, fouled spark plugs, carbon buildup in cylinders, and even shortened TBOs.

The compression check, a relatively simple test, is a true indication of the engine's ability to produce its rated power. Unlike its automotive counterpart, an aircraft engine compression check compares cylinder pressure against a known pressure, typically 80 psi. The FAA requires that the check be conducted by, or under the supervision of, a qualified mechanic for several reasons, not the least of which is the potential danger of being struck by the prop when pressure is introduced.

All cylinders will have some air leakage, therefore you will never get the perfect reading of 80/80; the question is how much leakage and where it occurs. Readings of 75/80 indicate a pretty healthy engine. Note the compression ratio numbers written on the cylinders in Figure 4-3. It is a common practice to list the variable number, which is always "over 80" for future reference. The FAA says leakage of 25 percent or more of the input pressure means trouble. Because the industry standard input pressure is 80, that means 60/80 and lower is cause for concern. There are three places where the air can leak past: the intake valve, the exhaust valve, and the piston into the crankcase. By listening at the air-induction inlet, exhaust pipe, and crankcase breather cap, you can determine which of the three is the culprit by the sound of rushing air.

In all three cases, however, there still is hope. Bad valves could be the result of a bit of carbon that is preventing complete closure. The mechanic, by giving the rocker valve an educated whack called "staking," hopefully can dislodge the carbon. A subsequent normal reading means you have just saved a bundle of money, but continued low readings indicate a bad valve.

In the case of air blowing by the piston, the problem may be that the engine has cooled too much and the oil has drained away from the cylinder wall, which reduces the airtight integrity. Running up the engine again, then performing another compression check, may make the difference. If there is no change in the pressure reading, you have a real problem. Readings for all cylinders should be within 5 psi of each other, indicating generally uniform wear. Readings below 60/80 on all cylinders, or more than 5-psi difference between one cylinder and the rest, require further investigation.

There is one last line of defense before actual removal of the cylinders. Your mechanic can use an instrument called a *borescope*, which permits an inexpensive, visual inspection of cylinder walls, rings, and the top of the piston without disassembly of the cylinder.

The 50-hour inspection—another good investment of time and money—may be performed by the pilot. It calls for a thorough preflight inspection and then a careful security check of all visible systems such as ignition, fuel/induction, cooling, lubrication, and exhaust. The main thrust is to assure that everything is tight, free of damage, and leak-free with no excessive wear or indications of heat damage. Finally, the oil should be drained and replaced. Oil loses effectiveness with time, making oil changes and clean oil filters extremely important to engine health and longevity.

In many cases, the history of an airplane is as important as its symptoms. For instance, if there has been a steadily increasing magneto drop over time, the problem could be old plugs, but a sudden increase could indicate ignition harness or magneto trouble.

Similarly, a steady increase in oil consumption over a long period is a sign of normal wear leading slowly toward an engine overhaul. If, on the other hand, there is a sudden, dramatic increase, that points to a more pressing problem. Hopefully, a visual inspection will turn up an oil leak that is easily fixed. If not, do a compression check for valve, ring, or valve-seat trouble. You might even suspect improper mixing of two different types of oil.

Normal wear over time increases tolerances between moving parts, so vibration and noise eventually will appear. Such irritants, if accompanied by a significant increase in oil consumption, may indicate that time for overhaul is at hand. Localizing such noises and vibrations can be of great help to the mechanic. In addition, it is helpful to know under what speed, power, and aerodynamic configuration the noise and vibration occur.

Not only is the health history of the engine important, but so is the operational history. It is of immense value to the mechanic to know how the airplane has been operated. Examples of operating conditions that will have a significant effect on the life expectancy of the engine are excessive high-power operations such as a towplane; regular operation from unimproved surfaces, particularly in dusty conditions; and any nonstandard operating procedures such as routine takeoffs with less than rated power.

Every airplane owner eventually will be faced with the question of overhaul. As your engine approaches TBO, there are several options, all with very predictable results. It is worth noting that no airplane escapes this "moment of truth," and the owner who flies "cheap" without an hourly allotment for engine replacement will have a far greater moment of truth than the one who has set money aside for the contingency. In addition to operating the airplane in accordance with the POH, the prudent owner will conduct 50- and 100-hour inspections, which have been proven to extend TBO and cut down costs in the long run.

Magneto

Preventive maintenance should include a timing check at 100 hours or annual inspection, whichever comes first. Routine maintenance typically is at 500 hours. At that time, the mechanic checks, among other things, contact point assembly burning and wear, distributor gear carbon-brush wear, cracking and chipping, and integrity of the impulse coupling shell and hub. A magneto always should be overhauled (or replaced) at engine overhaul or any other time conditions warrant it.

Cooling Systems

For the most part, preventive maintenance of the cooling system couldn't be simpler. Routinely check system integrity, with emphasis on keeping the entire airflow path free of obstructions. Also, routinely check the structural integrity of the baffles and fins.

Touch them, wiggle them if you can, to assure that the baffles are not only in good condition, but also are properly aligned.

One method of making sure the baffles are doing their job is to spray paint primer on the inside of the cowling over the baffles where the seals touch the cowl. Every 100 hours, check the inside of the cowling for rub marks where you have spray-painted. A lack of marks indicates that the seals are not contacting the cowling and air is flowing around the baffle instead of being properly channeled to cool the cylinders, which will translate into shortened TBO. Perhaps the most important thing to remember about this system is that no system is so simple that it cannot cause a problem if neglected.

5
Lubricating Systems

LUBRICATING FLUIDS, NAMELY OIL, SERVE MANY VARIED PURPOSES in reciprocating engines. One of the main purposes of oil is to reduce friction—to fill in the microscopic peaks and valleys on the surface of metal. Oil holds metal surfaces apart so the relative movement is actually between two layers of oil. This sliding effect greatly reduces friction and extends the life of the metals, so it is easy to understand why keeping your oil healthy has a significant impact on the health of an engine.

Viscosity is the most important characteristic of a lubricant. Viscosity is the oil's internal resistance to flow. It relates to how "heavy" or "thick" the oil is. Water, for example, has a very low viscosity; molasses is very high. The lubricant industry now uses kinematic viscosity as the standard measurement. Kinematic viscosity, measured in centistokes (cSt), is the measurement of flow time between two determined points. The instrument used to determine the rating is called a *viscometer*.

The first industrywide standards for piston-powered aircraft engine oil have been issued by The Society of Automotive Engineers (SAE). The new system defines specifications for all grades of oil including multiviscosity oils. Previously, military specifications were commonly accepted performance standards for civil aviation, but they did not define newer grades of oil such as SAE 30, SAE 50, and multigrade oils. The chart lists each SAE oil grade and the existing military specification. The new terminology for oils that meet the new SAE standards is also noted for both AllMineral and Ashless Dispersant oils.

The clearance between moving parts will determine what viscosity oil should be used. Proper viscosity assures that the oil won't separate, which helps reduce excessive friction. Other important considerations are pour point and flash point.

The colder the temperature, the stiffer the oil gets. Pour point is the coldest temperature at which the oil will continue to pour. Flash point is the coldest temperature that will still permit a momentary flash without sustaining combustion when a small flame is put next to the surface of the oil.

Oil also acts as a coolant. As oil comes in contact with high-temperature engine parts near the combustion chambers, heat transfers to the oil, which in turn transfers the heat to the outside air as the oil travels through the oil cooler. Friction itself also causes heat, as any roller-blader will tell you after sliding face down on a sidewalk. By bathing or splashing oil on moving parts where there is friction, the oil helps carry away the heat as well as reduce the friction between parts.

A third purpose of oil is its ability to cleanse. It gathers up particulate matter (such as water, dirt, dust, and flakes of metal and carbon) as it travels through the engine and holds them in suspension. Eventually the oil encounters the system filter, which traps the contaminants but allows the filtered oil to reenter the cycle through the engine.

The fourth function of oil is to prevent rust and corrosion. As an engine cools after use, moisture condenses onto the cylinder walls and other engine parts. This moisture, and other contaminants, lead to internal engine rust and corrosion. Oil coats the surfaces, thus preventing moisture and contaminants from actually contacting them.

The last major use of oil is to seal and cushion. Oil helps the piston rings form a seal against the cylinder wall, permitting maximum compression within the cylinder. It also cushions the shock of the moving parts.

TYPES OF OIL

Mineral oil, in one form or another, has been in use for years; the Wright Brothers used "A-Mobiloil" mineral oil in their early motors. Even though today's mineral oil, meeting military specification (MilSpec) MIL-L-6082B, is a well-established, common aircraft engine lubricant, it does have some important drawbacks. When aerated at high temperatures, especially after engine shutdown, oxidation takes place, which is the formation of carbon deposits. Even at temperatures of 150 degrees and lower, the combination of water vapor, lead compounds, and partially burned fuel tend to "cook" into sludge. This gooey mass clogs filters and can even damage engine bearings.

Some time ago there was a short-lived metallic ash detergent oil. It was a mineral oil with an ash-forming additive—metallic salts of barium and calcium. Initially, it appeared to be the answer to problems associated with mineral oil. It decreased the tendency for oxidation, reduced spark plug fouling to a minimum, lowered the tendency for preignition, and had minimal effect on the combustion process while it facilitated engine-cleaning action. The latter was considered especially noteworthy because as the oil traveled through the engine it removed carbon deposits and sludge. Unfortunately, metallic ash detergent oil was a disaster disguised as a blessing. The loosened deposits ended up clogging filters and oil passages and generally causing mayhem within the engine. Metallic ash detergent oil is no longer used in aircraft engines.

On the other hand, ashless-dispersant (AD) oil, meeting MilSpec MIL-L-22851, has practically taken over the reciprocating engine aircraft market. By using a nonmetallic polymeric additive, AD oil has done away with the carbon-forming problems of mineral oil without adding the ash-deposit problem of detergent oil. The dispersant additive causes particulate matter to repel one another, preventing sludge. At the same time, the dispersant holds the separated matter in suspension until it is removed when the oil is changed. Originally there was concern that the free-floating particles would act as an abrasive, forming a sort of flowing sandpaper that would wear out parts as it flowed by them. Experience has shown it to be quite the contrary. AD is such a good lubricant that many manufacturers require a new engine break-in period using mineral oil.

Synthetic Oils

With reciprocating engines operating at higher temperatures than ever before, yet being subjected to varying environments, new types of lubricants were proposed to satisfy the new needs. Synthetic oil is an attempt to solve the problem of large temperature variations. For instance, a synthetic oil can have the same viscosity at –20 degrees F as a non-synthetic AD oil at zero degrees F. Because synthetic oil has a lower internal friction than petroleum-based oils, it has excellent lubricating qualities at very low temperatures. In fact, the observant pilot would notice a 3–5 psi lower operating pressure than petroleum oil.

Engines using synthetic oil have started without preheat in temperatures as low as –40 degrees F. While there is definitely the potential to all but eliminate preheat, most manufacturers still recommend preheating the engine in extremely low temperatures. Nevertheless, it certainly means greatly reduced preheats, and perhaps the best part is no longer having to drain the oil just because the climate changes. Synthetic oil is honestly an all-weather oil. There are additional advantages to synthetic oil.

Synthetic oil allows a longer time between oil changes because it produces less oxidation at high temperature and has better wear characteristics than straight mineral oil. Probably the best benefit for the occasional pilot/owner is it adheres to metal better than other types of oil—certainly for weeks, and even months! This translates into longer engine life because it protects the cylinder walls from corrosion and provides instant lubrication on start-up, even for planes that aren't flown regularly. There are disadvantages, though. It has a strong tendency to soften rubber and resin products, so you have to be very careful about spillage and leakage. It is also much more expensive than the other types of oil, and while the extended oil change period tends to compensate for the added expense, a leaky oil system can literally run into big money.

It is a common misconception that you can't mix different brands of oil. Within the basic categories, all oils are compatible. All ashless dispersants meeting MIL-L-22851 are compatible with each other; they are also compatible with straight mineral oil. If, however, the engine is high time and has always used straight mineral oil, changing to

AD oil may not be as effective as with a lower-time engine. If you are planning to switch to a synthetic from either AD or straight mineral oil, then you should drain and flush the system as per manufacturer's recommendation. If you are using synthetic oil in your engine and need to add a quart or two of oil but don't have synthetic available, it is safe to use mineral oil. Bear in mind, as you dilute the synthetic with the mineral oil, you're defeating the purpose of using synthetic oil in the first place.

AIRCRAFT OIL SYSTEMS

Modern, light aircraft use a wet sump system similar to that shown in Figure 5-1. Oil is stored in the sump of the engine and is drawn out through a suction tube by the oil pump. The positive displacement, gear-type pump diagrammed in Figure 5-2 is the most common in light aircraft. Each time the engine-driven pump rotates, a fixed amount of oil is moved. A pressure relief valve maintains a constant system pressure as the pump speed varies. The pump has two spur gears meshed together. One is driven by the engine; the other follows. At the inlet side of the pump the teeth unmesh, causing the cavity volume to increase. This draws oil into the pump, where it fills the spaces between the teeth and is carried around. At the outlet side, the gears mesh, causing cavity volume to decrease, forcing the oil out of the pump. Here, in the close quarters of the meshing teeth, is one area where metal chips and other oil contamination can lead to trouble. The source for the oil pressure gauge is tapped off the pump outlet. To prevent gauge fluctuation and minimize oil loss if the line is broken, the hole is very small (approximately 3/16ths of an inch). The potential to clog such a small hole, or most oil passages, with sludge and other particulate matter is high, so a filtration system is employed.

In addition to clogging oil passages, solid contaminants and sludge can cause significant wear and damage to bearings, rings, cylinder walls, and pump vanes. Typically, a full flow filter is used, forcing all oil to pass through the filter each time it circulates. If only we had something like that filtering our arteries; think of the chocolate sundaes and cheeseburgers you could consume! The most common filter used in general aviation is the semidepth, which is a long, pleated sheet of resin-impregnated fibers, as shown in Figure 5-3. This sheet is rolled up around a steel core and is either put inside a metal spin on container or into a housing integral to the engine. To prevent oil system failure should the filter become clogged and prevent the normal flow of oil, a pressure-sensitive bypass valve is installed that will reroute oil around the filter. It is worth noting that the pilot will have no indication that the filter is being bypassed; contaminated oil will continue to flow through the engine until the next oil change or until sufficient damage is done to draw attention to itself. This is one of the major reasons why it is critical to change the oil and filter routinely.

A spring-loaded relief valve, downstream from the pump, is used to maintain constant system pressure as the pump speed varies with the engine. If the pump outlet pressure is less than spring pressure, oil continues through the system; if it is greater, the spring is displaced and oil is rerouted back to the inlet side of the pump, causing system pressure to reduce. This process happens so rapidly that fluctuations are not noticeable on the pressure gauge. An adjustable screw varies relief valve spring tension to permit system pressure calibration.

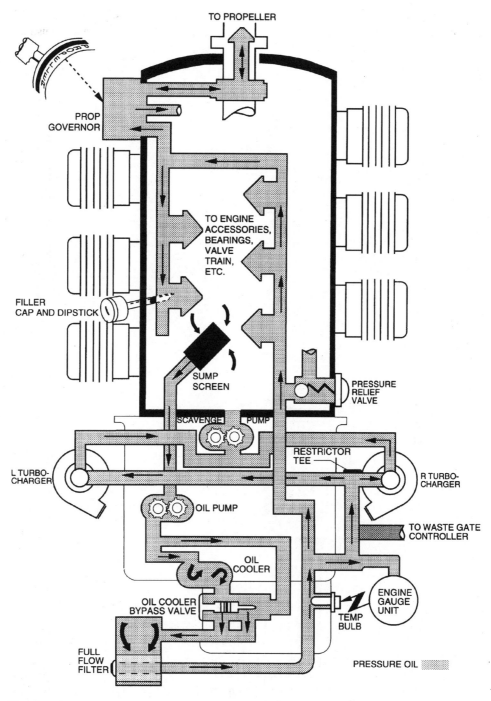

Fig. 5-1. *PA-46-350P Malibu engine oil system schematic.* (Courtesy of The New Piper Aircraft Company)

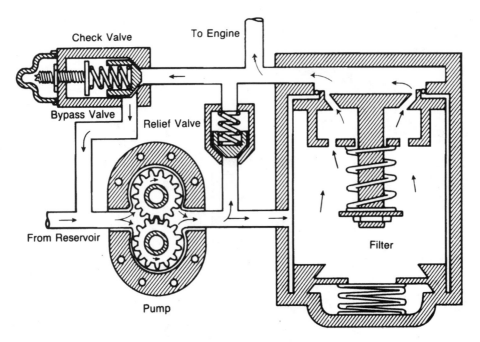

Fig. 5-2. *Constant displacement, engine-driven oil pump.*

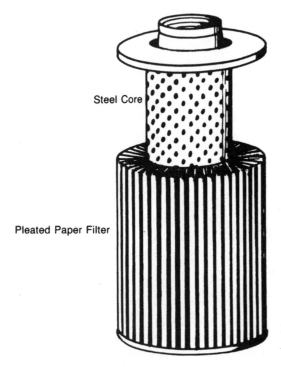

Fig. 5-3. *Aircraft oil filter.*

As oil travels through the engine, heat generated by combustion and friction is transferred to it. The oil must then be cooled before returning to the hot sections of the engine. An oil cooler, which is functionally similar to an automobile radiator, is used as a heat exchanger between hot oil and outside ram air. A thermostatic valve routes hot oil through the core of the heat exchanger for cooling or bypasses cold oil around the core. The sensor for the pilot's oil temperature gauge is located where the oil enters the engine after it has passed through or around the cooler. Some gauges are electrically powered, while others are mechanical.

Cooled oil is either sprayed or injected onto the crankshaft, camshaft, propshaft bearing, accessory drive bearings, cylinder walls, pistons, and various gears and parts. Both constant-speed propeller governors and turbochargers also use system oil. The oil then drains into the sump and the cycle begins again. To check oil quantity, a dipstick marked in quarts is provided; a filler tube is also provided to add oil. It is important to check the security of both the dipstick and the filler tube prior to every flight to assure that the caps are secure. Low ambient pressure during flight may cause oil to be siphoned out of the engine if either cap is missing.

PREVENTIVE MAINTENANCE

Maintaining an oil system is really a threefold consideration: selection of the proper lubricant, maintaining the required level of healthy oil, and regularly checking the health of the engine by having used oil tested.

It is critical that you only use the grade of oil specified by the engine manufacturer. If there is any question at all, call the manufacturer and ask. Using the wrong oil can not only drastically shorten the life of the engine but also can potentially lead to an in-flight catastrophic engine failure. Automotive oils, even when the numbers are the same, are not approved for aircraft use because they contain additives that will cause problems in aircraft engines.

Oil quantity should be maintained at the level recommended in the pilot's operating handbook (POH). This simple, pilot-performed preventive maintenance measure should be done before every flight. To assure accurate measurement, the aircraft should be in a level attitude. The dipstick should be removed, wiped with a clean rag, pushed all the way back into the dipstick tube, then removed again and read.

Cold-weather starting requires special consideration. Always follow the manufacturer's recommendations, but a good rule of thumb is to use a preheat whenever the temperature falls below freezing. It is very common for operators of aircraft in very cold climates to install winter oil baffles, which cut down on airflow cooling. When warm weather returns, however, be sure to remove the baffles.

In winter, excessive fuel priming before engine start washes oil from the cylinder walls with a twofold penalty: accelerated engine wear and potential fire hazard. Always keep the power setting low until the engine warms up, and avoid any abrupt power change prior to normal operating temperature. Oil, congealed by the cold, will sit in the bottom of the engine until the rising temperature thins it out enough for it to flow. Until that happens, the lack of oil flow results in excessive engine wear. Make no mistake;

starting an engine in excessively cold weather takes its toll on engine life. This is one area where synthetic oils, with their very low temperature flowability, have a significant impact on engine longevity. It is worth noting that even during summer months, engine oil still needs some warm-up time before flight.

It is important to change oil and filter at or before recommended times. The dirtier the atmosphere in which you operate, the more frequent the changes should be. There are two considerations that determine when to change oil: engine time and calendar time. It is common for operators to change engine oil every 25 or 50 hours, depending on manufacturer's recommendations. That's fine for aircraft that are frequently flown; however, many aircraft may not fly 50 hours a year. If an aircraft is an infrequent flier, you should consider calendar time. Oil should be changed at least three times per year regardless of engine hours. Oil in low-use aircraft will have a difficult time lubricating the engine because of fuel dilution and condensation.

Oil leaks shouldn't be taken for granted. Most pilots have come to accept oil stains on the underside of an airplane, but engines aren't supposed to leak oil. If there is a puddle under your airplane, or a long wet streak on the fuselage, find out why. If it isn't easily traceable, consult your mechanic. Don't ignore oil leaks; they don't go away by themselves. Knowing your airplane is one of the most important preventive maintenance tips. Every pilot should be familiar with the oil system's normal operating pressure and temperature; significant deviations are symptomatic of a problem and should not go unheeded. Similarly, never continue engine operation if either temperature or pressure exceeds red line; extensive engine damage follows closely behind! It is a very rare occasion when an engine fails without warning!

Another area where oil may show up inappropriately is on the spark plugs. During routine maintenance it is a good idea to check spark plugs for oil deposits. Oil isn't supposed to be able to get to that part of the engine, so you know you have a problem if there is more than just a trace. Bad piston rings are a likely culprit; they may be cracked or worn. This may not necessarily show up on a compression check either. If the compression rings are good, the check will be satisfactory, but the other rings may be bad, allowing oil to leak past. The only other route for oil to get to the plugs would be through worn valve guides.

An often-misunderstood principle is that engines that get little or no use have shorter life spans than engines that are used frequently. That may sound contrary to logic, but it is an established fact. And if you do start an engine, don't just start it; run it for a few minutes and shut it down! The engine case heats up, but before the oil really gets hot enough to boil out water and acid, the engine is shut down. As a result, moisture condenses on the inside of the engine, which leads to corrosion and rust. If you expect your engine to be inactive for a month or more, the engine should be pickled according to manufacturer's recommendations.

When changing oil, remove the filter and cut it open. Inspect the filter element for any contamination. A visible trace of metal in the filter is not always reason for concern. Some occasional metal flecks are normal, but excessive metal particles can indicate impending engine failure and should be taken seriously. Any traces of metal visually found in the filter should be sent along with the oil sample for determination of origin. To prevent inadvertent contamination of spin-on filters, which are sealed in a can, there is a special cutter that will remove the top without allowing any particles to fall into the filter itself. More than one pilot has been aghast to discover small chunks of metal in the filter, only to find out they were the result of cutting open the filter for inspection.

OIL ANALYSIS

One of the most effective preventive maintenance procedures you can do for the long-term health of your engine is to have your used oil sent away for spectrometric laboratory analysis. And while there are certainly those who feel it is a waste of time and money, many more swear by it. There are countless claims that the early detection of problems by oil analysis prevented what could have been catastrophic in-flight failures.

Labs that perform aviation oil analysis generally agree that about 75% of the samples they analyze indicate normal engine wear. Approximately 20% indicate some sort of abnormal wear, with 5% showing a critical engine defect that requires immediate attention. Considering that a critical engine defect could result in a catastrophic in-flight engine failure, it just makes sense to have your oil analyzed on a regular basis.

The reason for establishing a regular engine oil analysis program is because reciprocating engines are by design high-friction-causing machines with a lot of relative motion between moving parts like the crankshaft and crankcase or the connecting rod bearings. When two metals rub together, such as the piston rings against the cylinder walls, the resulting friction at the points of contact slowly wear away the metal, eventually leading to failure of the part. The metals used at the various contact points within an engine and oil system are known as "wear metal." Oil analysis labs know what wear metals are used in specific areas, and traces of a given metal will indicate what area or areas are experiencing wear. But there are more reasons than just wear to worry about.

Reciprocating engines attract a wide variety of nasty contaminants such as dirt, fuel, water, oxidation, nitration, and for those few liquid-cooled engines—glycol. Add these to particles of metal floating in the oil, and the oil's effectiveness is dramatically eroded. The problem is these contaminants are usually microscopic and can't be seen by the human eye, so just looking at oil when you change it, or feeling it with your fingers, is of little value.

Oil analysis is a trend-indicating program requiring a minimum of three sample periods to establish a meaningful trend. It is not a one-time or occasional thing to do. Therefore, it is suggested that oil samples be taken every 50 operating hours or 3 months, whichever comes first. At the very minimum, for the program to be effective, samples should be taken with every oil change.

It is important to note that there is variation among labs in their methods of recording data, so you should pick one lab and stick with it over time. This assures the structure of using the same system and builds up historical data for long-term comparison. Another important aspect of the analysis is to completely and accurately fill out the questionnaire. Being completely candid when answering questions about operating conditions and procedures helps the analyst to make recommendations based on what the oil analysis results indicate. Remember that a given type of metal may be used in several different areas of an engine and oil system. Knowing operational information helps the analyst interpret the data.

Two-way communication with the analyst is crucial. For instance, if the analyst makes a recommendation to look at a specific part based on the results and it turns out nothing is wrong with that part, there is still a problem somewhere! Call the analyst and explain that you followed the recommendations and nothing seemed out of the ordinary. There will then be a review of the data, and based on that information, further recommendations will be made.

Most labs run two types of oil analysis tests on a given sample. They are a spectrometric test and a physical property test. A spectrometric test is one in which the sample is subjected to an extremely high-voltage electrical energy that causes the elements in the oil to give off signature colors and brightness. By understanding these signatures, analysts can interpret how much of a given element, typically reported in parts per million, is in the sample. Analysts also know that specific metals are used to manufacture specific engine parts, an abbreviated version of which is in Table 5-1. For instance, copper is used in wrist-pin bushings and cam bushings, so determination of the origin of the metal contaminant is often possible.

The physical property test checks for such things as the presence of water, fuel dilution, oxidation, and viscosity, which is the measure of a lubricant's resistance to flow. Viscosity increases with time in service due to such things as oxidation, picking up contaminants in the oil, overheating of the oil, improper air/fuel ratio, and an excessive increase in solid materials trapped in the oil. It is also possible for there to be a decrease in oil viscosity due to a bad fuel-injector tip causing fuel dilution of the oil.

The physical properties test will also check for total solids. A sample of the oil is diluted with a solvent then spun in a centrifuge, causing the larger solids to settle, which then makes them easily and accurately measurable. One of the results of this portion of the test is the discovery of carbon particles indicating that there has been an incomplete burning of the fuel, which is a good indicator of the engine's combustion efficiency.

When drawing the oil sample you want to make sure the oil is hot. There are two locations from which you can draw a sample: the dipstick port and the engine sump. You would draw a sample from the dipstick port when you suspect a problem and will be doing tests more frequently than scheduled oil changes. The more common location would be the oil sump during an oil change.

There are many benefits to engine oil analysis. For instance, it should play a major role in breaking in new or recently overhauled engines. Much of an engine's break-in actually occurs during the first hour. The trace amount of copper, chrome, aluminum, and iron reflect how the break-in went and whether or not the engine is likely to make it to TBO. When breaking in either a new or overhauled engine, the oil should be changed frequently. There is a high wear rate during the break-in period, and metal particles can imbed in bearings and severely shorten the engine's life. Typically, oil should be changed at 5 hours, or after the first flight, then again at 10–12 hours, and again at approximately 25 hours. It is a good idea to obtain oil samples at each change, plus at the 60-hour point. If you are using synthetic oil, the 10-hour analysis will show significantly lower levels of

Table 5-1. Source of Oil Contamination

Silicon	A measure of airborne dust and dirt contamination, it usually indicates improper air cleaner service. Excessive dirt and abrasives accelerate engine wear and can greatly increase operating costs.
Iron	Indicates wear originating from any and all steel components, such as cylinder walls, rings, shafts, splines, gears, etc. High iron content can indicate corrosion if the engine has an inactive history. Often it will clean out with regular usage, if cylinders, cam, and lifters are not pitted.
Copper	Indicates wear from bearings and/or bushings.
Aluminum	Indicates piston metal, piston pin plugs, and can confirm airborne dirt.
Chromium	Originates from wear of engine parts that have been chromed, primarily compression rings or cylinder walls.
Magnesium	Water reacts with magnesium casings and is carried in the oil. Magnesium also may be an oil additive.
Silver	Present in the bearing alloys of a limited number of engines, such as the Lycoming supercharged engine, radial-engine master rods, and E series continentals front main bearing.
Nickel	Can indicate wear from certain types of rings, bearings, and valves or turbo shaft.
Tin	Indicator of wear from bearings.
Lead	In gasoline engines, the main source of lead is from tetraethyl lead contamination.

trace metals than if you had used mineral oil. This is because synthetics result in less wear due to their higher film-temperature gradient.

The overall oil analysis program benefits include reduced in-service equipment failures, easier scheduling of repairs, identification of maintenance deficiencies and operator abuse, simplification of the decision to "buy or repair," and help in the evaluation of trade-ins and rentals.

To assure engine longevity, the prudent pilot will use the appropriate oil for the engine and operating conditions; maintain sufficient quantity; change it in accordance with manufacturer's recommendations; maintain an engine oil analysis program; and be aware of what engine instruments are saying. If pressures and temperatures are running at a significantly different level than normal, there is something wrong.

6
Fuel Systems

UNDERSTANDING THE FUEL SYSTEM OF THE AIRCRAFT YOU ARE going to fly is probably one of the most critical aspects of safe flight. Accident statistics have fairly consistently shown that approximately 25% of all accidents are fuel mismanagement related. When it comes to fuel management, assumptions can be deadly, and that starts with the all-too-common assumption that the performance charts are accurate.

Manufacturers don't lie in the development of performance charts, but they do stack the deck in their favor. The aircraft they use in the development of those charts is in absolutely perfect condition in every aspect, but they don't stop there. The vents are sealed shut on the outside because airflow diverted through air intakes causes drag. Only the antennae required by aircraft certification are sticking out of the airframe to create their own special brand of drag. There are no steps to help the pilot get in the airplane, and the CG is as far aft as legally permitted. You get the idea; there's no way you can possibly duplicate the performance displayed on those charts.

Beyond performance, however, comes a host of system-related issues that can cause an abrupt termination of a flight. It is impossible to overstate the importance of not only studying the POH's information about how the fuel system works, but also any options that may have been added, such as extended-range wing tip tanks. To illustrate the problem, some tip tanks feed fuel to the main tank when you activate an electric tip tank fuel pump. More than one pilot has turned on all fuel pumps prior to takeoff and simply left them on, which in some aircraft may be an appropriate procedure. In this case, the fuel in the tip tanks will be forced into a full main tank, causing main tank fuel to vent overboard through the fuel vent. The result would be fuel dumping and a potentially dramatic decrease in endurance performance. It is important to know as much about fuel and your fuel system as possible, so let's start with the basics.

CHEMISTRY OF COMBUSTION

It is a common belief that a few million years ago T-Rex, Godzilla, and a bunch of other dinosaurs died when an asteroid hit earth and caused a new Ice Age. They got buried, rotted, mystically turned into petroleum, and we now pump them into our gas tanks, which lends precious little credence to one gas company's slogan: "Put a tiger in your tank." Contrary to what most people believe, however, the jury is still out on where petroleum came from. One thing is for sure: The dark, greenish-brown viscous fluid contains many hydrocarbons, and hydrocarbons mean fuel.

There are three primary classifications of fuel: solids, gases, and liquids. Let's face it; flying would lose a lot of its appeal if you had to shovel coal while doing it. And while there have been some earnest experiments in using gaseous fuels such as propane, of the three, the one best suited for use in aircraft are liquid fuels. Avgas (aviation gasoline) is one of a number of hydrocarbon compounds obtained from petroleum. There are actually quite a few methods of producing gasoline from crude oil, but the most commonly used are fractional distillation and thermal cracking.

Fractional Distillation

Fractional distillation is a process whereby the crude petroleum is heated, vaporized, then condensed (Figure 6-1). A specific hydrocarbon liquid—for instance, gasoline—will vaporize at a given temperature. So when the crude petroleum is heated and held at that temperature, the liquid that condenses out will be gasoline. When all of the gasoline has been distilled out, the petroleum temperature can be raised and a different, heavier hydrocarbon liquid such as kerosene will condense out, and so on.

During the relatively short period since the discovery and use of crude petroleum began, many of the less sweet (very clean) oil reserves have been depleted. That is not to say that there isn't still a lot of crude oil available throughout the world, but the quality, type, and quantity of the fractions that can be obtained from different sources of crude differ. Refiners naturally prefer to use crude that will give the greatest yield for the least cost. Since the cost of operating a refinery tends to increase with inflation, as time goes on the remaining, less productive crude will naturally yield more costly products.

Even though fractional distillation appears to be a simple process, it is actually relatively expensive because it is not very efficient at extracting, for instance, all of the possible gasoline in a given volume of crude petroleum.

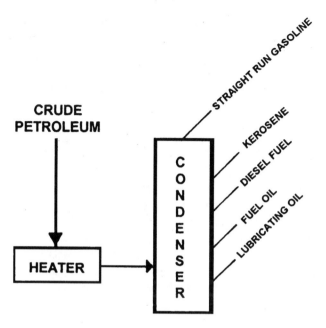

Fig. 6-1. *Fractional-distillation process.*

Thermal Cracking

Thermal cracking, on the other hand, is a process that will substantially increase the amount of gasoline extracted from the crude petroleum after it has completed the fractional distillation process. What remains are the heavier carbon liquids that are not only unsuitable for gasoline, but even for lubricating oil. This remaining crude oil is put in a pressurized vessel and heated. This process actually forces a chemical change called "cracking," causing some of those heavier carbon liquids to turn into "cracked" gasoline. The combined fractional distillation and thermal cracking processes produce an increased amount of gasoline obtained from a given volume of crude petroleum, which drives down the overall cost of production per gallon.

Aviation Gasoline

Of all the types of fuel available, the one we call avgas meets the myriad needs of aircraft piston engines. It has a high energy content and vaporizes easily throughout the wide, ambient operating range of an aircraft engine while not being so volatile that it is likely to cause fuel vapor lock. It also has a sufficiently high octane rating to allow its use with high-compression engines without the problem of detonation. Finally, its chemical composition is such that it minimizes the chance of gumming up or corroding the engine when used.

Octane Rating

Gasoline is used in a wide range of applications, from lawn mowers to high-altitude reciprocating engine-powered aircraft. No one type of gasoline can meet such a vast array of requirements. A method established early on to classify gasoline is by its antiknock value, called octane.

Two types of hydrocarbons found in gasoline are isooctane and normal heptane. Isooctane has a high antiknock value, while normal heptane does not. Octane rating is simply the percentage of isooctane mixed into normal heptane. For instance, 80-octane gasoline is a solution of 80% isooctane in normal heptane. The higher the octane number, the greater the antiknock capability.

As new metal alloys were perfected, they permitted the development of engines capable of greater manifold pressures, increased cylinder compression ratios, and significantly higher power output. Before long, engines were burning 100-octane avgas, pure isooctane, and engineers were still pushing engines and causing them to succumb to fuel knock.

You may recall from the chapter on powerplants that an engine is most susceptible to engine knock, called detonation, at high power settings. This is particularly true if combined with improper leaning. Excessive temperature can cause the fuel/air mixture within the engine's cylinders to detonate explosively. The same problem exists by increasing cylinder pressure because it results in an increase in temperature. So there began to exist a need to push avgas beyond 100 octane, and the result was the development of fuel performance numbers.

Performance Numbers

It was discovered that by adding tetraethyl lead to avgas it was possible to increase its antiknock value in excess of 100. It is also true that the antiknock value of a given fuel will vary with the fuel/air ratio, so a performance number system was developed that uses two numbers. Take, for instance, 100/130 avgas.

The number 100 in 100/130 avgas represents the performance number at the engine's leanest fuel/air ratio. The 130 represents the performance number at the engine's richest ratio. These fuels opened up new horizons for engine manufacturers until it was discovered that tetraethyl lead has a number of drawbacks, including being a health hazard. Challenged again, manufacturers began developing low lead versions of some types of avgas through the use of other chemicals. The most common types of avgas used in general aviation today are 80/87, 100/130, and 100/130LL (low lead).

The bottom line regarding avgas is that when refueling an aircraft, never use a grade of fuel lower than what is recommended by the manufacturer. In general, you don't want to use a higher grade if you can avoid it, although the use of avgas one grade higher is acceptable if you have no choice.

GRAVITY FEED VERSUS PRESSURE SYSTEMS

The purpose of a light aircraft fuel system is to store fuel safely and deliver the correct amount of it at a uniform flow to the carburetor or other fuel control unit. There are two types of systems: gravity feed and pressure.

Gravity-feed systems, such as shown in Figure 6-2, rely on the force of gravity to deliver fuel from the tank to the carburetor, which limits them to high-wing aircraft. The Cessna 152 uses a gravity-feed system because it is very simple, relatively inexpensive, and does not require a fuel pump. This type of system is limited to smaller, single-engine aircraft.

The most significant drawback to the gravity-feed system is its tendency to develop *vapor lock*, a condition where fuel changes from a liquid to a vapor. The fuel lines, filled with vapor, are unable to supply sufficient fuel to maintain engine combustion, resulting in a fuel-starved engine. Vapor lock is caused by excessive fuel temperature, high-altitude operations, or a combination of the two. Fuel vaporization is most likely the result of shutting down an engine on a hot summer day, causing fuel lines, located under the cowling and next to the engine, to be exposed to extremely hot temperatures with no cooling airflow. High-altitude operation, where there is lower atmospheric pressure, also may induce vaporization at a lower temperature. The solution to the problem of vaporization is to provide positive fuel pressure with a fuel pump.

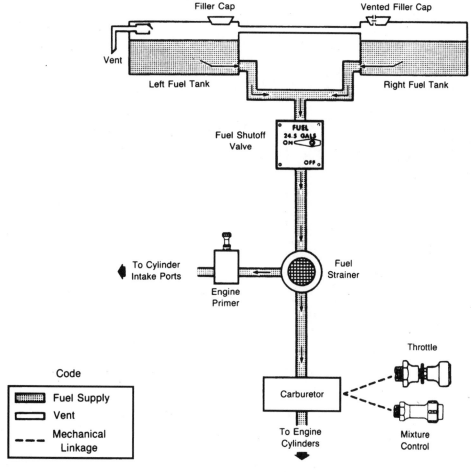

Fig. 6-2. *Cessna 152 gravity feed fuel system.*

Chapter Six

In the pressure system, the engine-driven pump draws fuel from tanks located anywhere in the airplane and discharges it under positive pressure to the carburetor. This permits greater flexibility in tank utilization and minimizes vapor lock potential. A second or auxiliary fuel pump is used for priming, engine start, and as a backup in case of engine-driven pump failure.

As aircraft size increases, so does fuel system complexity. More sophisticated than the Cessna 152, the Beech M35 Bonanza is designed for longer trips and instrument flying. Its pressure-feed fuel system, shown in Figure 6-3, is somewhat more complicated than the C152's gravity-feed system, plus it has auxiliary fuel tanks.

When a pilot transitions from a single to a twin-engine aircraft, the increase in fuel system complexity becomes readily apparent. Multiengine aircraft are designed for long-range, instrument-flight, and emergency single-engine operation. The Beech B55 Baron, depicted in Figure 6-4, has a more complex fuel tank and fuel/engine selector arrange-

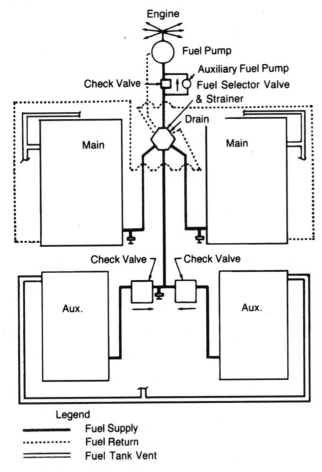

Fig. 6-3. *Beech Bonanza K35 and M35 fuel system schematic.*

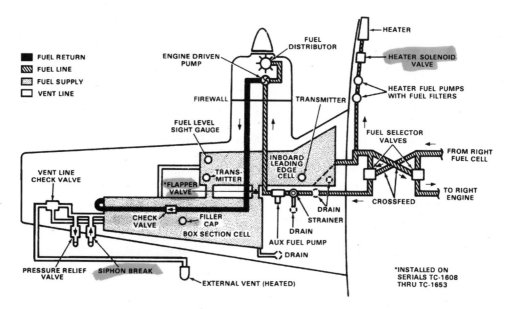

Fig. 6-4. *Beech Baron B55 fuel system schematic.*

ment than virtually any single-engine airplane built. The drawback of increased fuel system complexity is the potential for increased fuel mismanagement. In some multiengine aircraft, for instance, it is possible to unintentionally cause an unstable balance situation by burning fuel from tanks in the wrong sequence. Each aircraft has its own peculiarities, and good operating practice dictates a thorough understanding of system procedures.

Fuel Tanks

Fuel tanks come in all shapes, sizes, and locations. Integral fuel tanks are those that are permanently built into each wing. Tip tanks can be attached to the wing tip and often contain auxiliary fuel, while fuselage tanks may be located almost anywhere in the body of the airplane.

Temporary, portable fuel tanks can be located in the cockpit to provide supplementary fuel for extended flights, such as ferrying aircraft over water, but they present a considerable fire and noxious gas hazard. All tanks must be protected from vibration, a principle cause of deterioration and leakage, and must be able to handle fuel expansion as a result of heat.

Another consideration is surging, which may cause a fuel interruption resulting from fuel flowing toward the wing tip during turning maneuvers. The problem can be reduced by placing baffles inside the fuel tank, which will slow lateral fuel movement. Tank design must also include a method for the pilot to check for fuel quantity and quality during the preflight.

Most fuel tanks are made of either preshaped, riveted aluminum alloy or synthetic rubber. The aluminum "wet-wing" tanks use a sealant along their seams to prevent fuel leakage. The problem is that the sealant will, over time, deteriorate, resulting in fuel leakage. While having avgas leaking out of the wing and running down the fuselage or onto the ground is a major safety problem for several reasons, there is yet another problem associated with seal deterioration.

As fuel tank sealant ages, pieces break off and can become grit or a powderlike substance, which will clog a fuel filter. Simply put, aging sealant can result in engine failure. Whether that occurs during taxi, on a takeoff roll, or enroute over the mountains is simply a matter of chance. Therefore, the wet-wing fuel tanks of aircraft that are more than 10–15 years old should be inspected for possible sealant deterioration.

One way to eliminate the sealant deterioration problem of wet-wing tanks is to weld the seams rather than seal them. As you might anticipate, this is a costly process that results in a heavy, expensive tank. And it simply opens you up to a different set of problems such as vibration fatigue. Another option is fuel-resistant, synthetic rubber bladders, which are lighter than metal and very flexible. They are fitted easily into available space, relatively easy to replace, and much less susceptible to vibration fatigue. But even they have their problems.

Some types of rubber bladders become dry and brittle with age and begin to leak; however, filling the tank after each flight usually prevents this problem. Tanks made of Goodyear BTC-39 synthetic rubber (used in many Cessna, Beech, Rockwell International, and Piper aircraft during the 1960s and 1970s) developed a different problem— softness. An airworthiness directive (AD) required them to be inspected annually for deterioration; use of BTC-39 synthetic was finally discontinued by Goodyear, but if you own or regularly fly one of those aircraft, it is worth looking into.

Another problem that affects all rubber fuel bladders is wrinkles. When a synthetic rubber fuel bladder is installed, it is very difficult to get out all of the wrinkles. On the ground this acts as a water trap, but in-flight motion dislodges the water, causing a high potential for engine failure. This has become such a concern in the big Cessna singles (180, 182, 206, and 210) that the company issued an owner-advisory bulletin stating, among other things, that owners should "gently move and lower the tail to the ground" during preflight, which will hopefully cause any existing water to dislodge and show up in the preflight fuel sample. And remember, when you drain a fuel sump, a sample of a few ounces may not be enough to completely empty the fuel drain line and get to the water above it. Always take several samples from every sump.

That said, several experiments have shown that such a procedure may have little effect in removing entrapped water from these Cessna models. Consequently, AD 84-10-01 was issued requiring, among other measures, the installation of additional quick drains in these aircraft and an extensive check of the bladders for wrinkles and their effect on trapping water. If, after compliance with the AD, the bladder still traps more than three ounces of water, then an elaborate preflight inspection procedure is required, plus additional annual inspection considerations. It is simpler, and probably safer in the long run, to replace the fuel bladder with a new one.

Other parts of the fuel tank include vents and overflow drains. As the fuel level of a tank decreases, the fuel vent allows air to fill the space. It is important to preflight the vent. If a vent becomes blocked, fuel starvation will stop the engine and possibly collapse the tank. Overflow drains act as safety valves for fuel when it expands as a result of heat.

Nevertheless, the prudent pilot will allow for fuel expansion when refueling during the warmer months.

Fuel lines, normally made of annealed aluminum alloy or copper, must be of sufficient diameter to allow double the required flow rate at takeoff power. They must be protected from excessive vibration and, where actually connected to the engine or airframe, flexible hosing is required. Wherever visible, hoses should be checked for chafing during preflight. While manufacturers try to route fuel lines away from exhaust manifolds and other hot areas, it is not always possible. Asbestos tape serves as heat protection and should also be checked on preflight.

Fuel Contamination

Water can enter a fuel tank in three ways: The first way is by the formation of condensation on the inside of partially filled tanks when outside air temperature drops. Topping off tanks after each flight eliminates this possibility.

The second way is for water to leak past the fuel filler cap, a problem that exists in particular when the aircraft has the type of recessed filler caps that form a cup in the wing. While some caps are only slightly recessed, such as the one illustrated in Figure 6-5, any recess at all will collect rain. To minimize this problem, inspect filler caps, seals, and ports every preflight. Also check for fuel stains trailing behind the filler cap. In flight, reduced pressure over the wing causes fuel to stream out of a leaking filler neck. If you see fuel streaks behind the filler, not only are you losing fuel in flight, but also you are likely to be collecting water in your tank when the aircraft is on the ground in the rain.

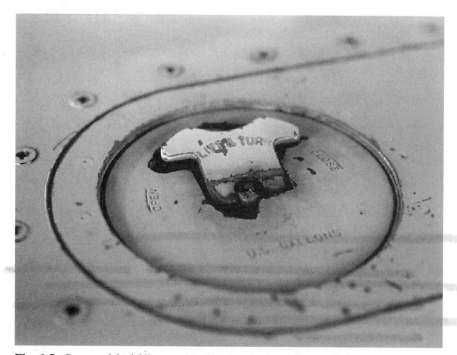

Fig. 6-5. *Recessed fuel filler cap in wing.* (Photo by author, courtesy of Frasca Air Services)

Chapter Six

The third way for water to enter a fuel tank is for it to be pumped in with the fuel. FBO fuel trucks and storage tanks suffer from the same problems as airplane fuel tanks about collecting water, but they have filtering systems built in. Reputable FBOs properly maintain this equipment and train personnel to use it correctly.

A strainer is located between the fuel source and carburetor to trap impurities and water. Quick drains usually are located at the lowest point in each fuel tank—and for the entire fuel system—where water and sediment tend to collect. During preflight, the pilot should check a fuel sample for purity and correct color. Any mixing of fuels will cause the color to become clear. Water, heavier than avgas, will sink to the bottom of the cup and form a distinct "bubble," while sediment will appear as floating specks. Always use a clear container for collecting the fuel sample so it may be inspected carefully.

Fuel Pump and Delivery System

Pressure systems have two types of fuel pumps: engine-driven and auxiliary. The engine-driven pump runs off the accessory panel at the rear of the engine. Typically an eccentric sliding vane type, it provides a positive displacement with a large volume output of fuel to the carburetor. As with most fuel pumps, the rotor is lubricated by the fuel itself, virtually eliminating the need for any kind of preventive maintenance.

The auxiliary fuel pump is required as a backup for the engine-driven pump. Originally these pumps were hand operated and known as *wobble pumps*. Today, auxiliary fuel pumps are powered by electric motors. In addition to their role as backup pumps, they are commonly used during engine start to build up fuel pressure, during takeoff as a safety margin, at altitudes above 10,000 feet to reduce potential for vapor lock, and during emergency operations.

In most light aircraft, a single pump is sufficient for normal operations, but in larger aircraft, electric boost pumps are often required. The purpose of the boost pump is to supply fuel under pressure to the engine-driven pump, thus preventing vapor lock. Failure of a boost pump usually means operational limitations imposed on the aircraft, including altitude restrictions and even a drastically reduced useful life of the engine-driven pump.

The purpose of the fuel selector valve is fuel manipulation by the pilot, including tank to engine, tank to tank, and (in multiengine aircraft) cross-feeding one engine from the opposite fuel tank. In the event of engine fire, emergency procedures call for the fuel to the engine to be shut off, a procedure that alone may put out the fire. As the pilot's options increase, so do the hazards, therefore several safe operating practices should be adopted.

First, never change the fuel selector position just prior to takeoff. Why switch to the unknown?

Second, never operate by feel alone. Always check visually; some fuel selectors require you to go through the OFF position when switching from one tank to another.

Third, always test all tanks while you still have options. Don't deplete one tank without knowing for sure that the other tank will work properly.

Fourth, never run a tank empty. You are creating an emergency condition when you do. It may take longer than anticipated for the fresh tank to recharge the system and get fuel to the engine, but—more importantly—airplanes with fuel injection are highly susceptible to the termination of fuel flow. Getting a fuel-injected engine started again after running a tank dry can be a major problem.

Finally, flight instructors should be aware that when they switch off the fuel selector to simulate an emergency, they have created one! There have been numerous forced landings when the fuel selector failed after being switched off. For a number of years, instructors changed their tactics and went to pulling the mixture control, but even that can result in an emergency if there is a control failure. A dead engine is a windmilling engine that creates staggering drag. Instead, go to zero thrust by pulling back the throttle; at least the engine is still running.

Both manual and electric primers supply a small amount of fuel to the cylinders for engine starting. The manual type is a single-action piston pump, which requires the pilot to pull out the primer, creating a partial pressure that causes fuel flow into the primer cylinder. When pushed in, the fuel is forced through a primer distributor to either individual cylinders or the intake manifold, depending on the type of system.

The main advantages of a manual primer are that it is inexpensive and simple. A significant disadvantage is the fire hazard that results from fuel routed through the cabin to the primer. During preflight, take a very close look at the primer itself, in front and behind the instrument panel, to make sure there is no fuel leak. Under the cowling, look for and check the condition of the fine primer tubing that runs into the cylinders or intake manifold.

The electric primer is a solenoid plunger. It acts like a door, permitting fuel to flow from a source pressurized by the auxiliary fuel pump to the primer distribution system. Because it is controlled by a remote switch in the cabin, you would preflight this system the same way you would a manual engine primer. The advantage is that there are no fuel lines running into the cabin, so you don't have to worry about a potential fuel leak.

The fuel pressure gauge takes its measurement at the carburetor inlet. There are two types: Bourdon tube and autosyn transmitter. The Bourdon tube is a small, coiled tube with a movable, sealed end. A fuel pressure increase causes the tube to unwind, and the end mechanically moves an indicator needle in the gauge. It has the advantage of being simple, inexpensive, and totally self-powered, but like the manual primer, it is a fire hazard because fuel is actually brought into the cabin to operate it.

The autosyn transmitter measures fuel pressure in the same manner as the Bourdon tube, but the pressure-sensing device is mounted on the engine side of the firewall with a transmitter armature electrically sending the appropriate indication to the cabin instrument. While it does require electrical power to operate, its significant advantage is that no fuel ever enters the cabin.

FUELING CONSIDERATIONS

As an aircraft moves through the air, the resulting friction causes it to become negatively charged, much the same as rubbing your shoe on a carpet. Cold, dry weather makes it worse. The earth, which tends to be positively charged, would cancel the airplane charge when it lands, but tires make very poor conductors of electricity. Along comes the flight line attendant driving the fuel truck.

Unless the aircraft is grounded to both the fuel truck and the earth, a spark will likely jump from the aircraft fuel tank to the hose nozzle, causing an explosion. While most pilots agree that proper grounding procedures should be followed when returning from a flight, many believe these procedures are unnecessary when refueling an aircraft that has not been flying. The very motion of the fuel flowing through the nozzle creates static electricity and may cause a spark between the aircraft and the refueler. For that reason alone, proper grounding procedures always must be adhered to when refueling.

Another refueling concern is contamination. This problem can largely be avoided by simply buying fuel from a high-volume, reliable dealer. Fuel that sits for a long time tends to collect various forms of contamination. And FBOs that obviously don't keep up their facilities are unlikely to be doing a very good job of maintaining their fueling system.

Prudent pilots will always assure that the aircraft is fueled properly by monitoring the fueling procedure and actually checking the fuel themselves. Don't insult the flight line attendant by being too obvious, or the quality of service may diminish rapidly. Instead, talk about the weather or sports, but keep one eye glued to what's happening at all times. Check and double-check fuel pumped in your aircraft; there have been instances where fuel trucks have been misfueled! To help prevent misfueling, have all refueling ports distinctly labeled with the fuel type and install misfueling-preventive hardware available for most aircraft. This two-part system of preventing incorrect fueling requires the installation of restrictors that decrease the size of the aircraft fuel tank opening. All new aircraft will come factory-equipped with the new fueling ports. The other part of the system, an oversized jet fuel hose spout, is being installed on refuelers by FBOs around the country.

Finally, never trust fuel gauges; they are notoriously inaccurate. You should purchase or build your own dipstick for your tanks and always check fuel quantity that way. Even when you tell a flight line attendant to top off the tanks, there is no guarantee that it will happen, despite the best efforts of the refueler. For one, unless the aircraft is sitting dead level on the ramp, the result can be a partially fueled tank. There are also nooks and crannies in most fuel tanks that will trap air. Wherever there is air, there isn't fuel. It's not a bad idea to shake wings after refueling, then top it off again if you truly want full tanks. The point is that you don't always get what you think, so take every possible precaution and build a fudge factor into all of your calculations.

TROUBLESHOOTING

If the engine turns over but simply refuses to start, the cause could simply be no fuel. As simplistic as it may seem, the tanks may be empty or the fuel valve off. Check both before calling your friendly mechanic in the middle of the night. Other, more complicated problems requiring a mechanic's expertise could include inoperative primer, a plugged or ruptured fuel line, an improperly set or inoperative carburetor mixture control, and any one of several carburetion problems.

Perhaps even more frustrating is an engine that roars to life only to die suddenly. If the airplane has been flown recently and it is a warm day, one of the most common causes is vapor lock. The pilot should follow the hot-start procedure outlined in the POH to purge the fuel lines of vapor. Other possible culprits include a clogged fuel vent line, which may be obvious upon inspection; an inoperative engine-driven fuel pump, which must be checked by a mechanic; a filled fuel strainer, easily cleaned by the pilot; or water in the fuel system, which the pilot can usually drain.

If the engine runs but excessive black smoke is coming from the exhaust, you need a mechanic. Probable causes are the idle or cruise (depending on where it happens) engine mixture setting being too rich; a continuous primer leak into the intake manifold; or any one of several other kinds of carburetor problems. Finally, if the engine starts and responds appropriately as you increase the throttle but doesn't seem to develop enough power to make the aircraft move forward, shut down the engine, get out, and remove the tiedown chains.

7

Turbocharger Systems

MOST PILOTS KNOW THAT A NORMALLY ASPIRATED ENGINE LOSES power as the aircraft climbs because air density (the number of oxygen molecules per cubic foot) decreases as altitude increases. Similarly, high ambient air temperatures cause an engine to produce less power, again due to decreased air density because warming the air causes the oxygen molecules to spread apart, lowering the density. So, if engine power is directly proportional to the mass of air and fuel burned in its combustion chambers, it should be a simple matter of increasing the flow of fuel/air to regain lost power.

In a way, it really is that simple. A larger fuel pump can increase the flow of fuel and an air pump (compressor) can pack more molecules into the cylinders. The complexity comes in constructing, attaching, and driving the compressor.

SYSTEM OVERVIEW

Mechanically driven compressors, called superchargers for their act of increasing the density within a specific volume, have been with us nearly as long as the airplane itself. At one time, the geared supercharger equipped virtually every transport or military aircraft. Most big radial engines have them, though they require a lot of power to drive. You still can find examples of mechanically driven compressors on diesel trucks and drag-racing cars. Aircraft, on the other hand, use exhaust-driven turbo-superchargers such as the Piper Malibu Mirage turbo-induction systems shown in Figure 7-1.

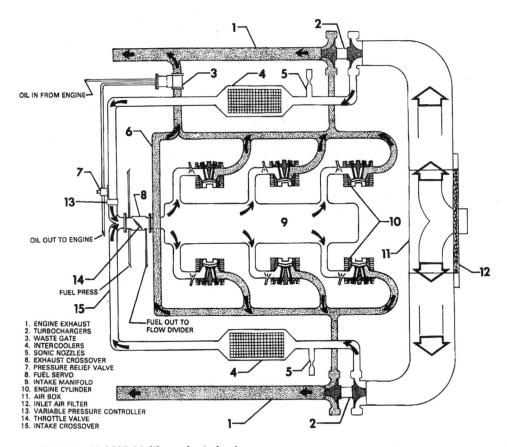

OIL IN FROM ENGINE

OIL OUT TO ENGINE

7
13
8

14
FUEL PRESS
15

FUEL OUT TO
FLOW DIVIDER

1. ENGINE EXHAUST
2. TURBOCHARGERS
3. WASTE GATE
4. INTERCOOLERS
5. SONIC NOZZLES
6. EXHAUST CROSSOVER
7. PRESSURE RELIEF VALVE
8. FUEL SERVO
9. INTAKE MANIFOLD
10. ENGINE CYLINDER
11. AIR BOX
12. INLET AIR FILTER
13. VARIABLE PRESSURE CONTROLLER
14. THROTTLE VALVE
15. INTAKE CROSSOVER

Fig. 7-1. *PA-46-350P Malibu turbo-induction system.* (Courtesy of The New Piper Aircraft Company)

Turbo-supercharging (generally referred to as "turbocharging," an operation performed by "turbochargers") also has been around for a long time. The first experiments with turbochargers on aircraft took place in the early 1920s. Some of the more advanced World War II military airplanes had them, notably the P-47 Thunderbolt and P-38 Lightning fighters, and B-17 bombers—all of which could top 30,000 feet! Their advantages are many: light weight, compact, simple installation, and a "free" source of power.

Maintaining Sea-Level Power

In general aircraft use, the turbocharger's role is not to increase engine power at sea level, rather it is designed to maintain sea-level power as the aircraft climbs. That's an important distinction because a turbocharger has the ability to overboost an engine at sea level, which can result in serious engine damage.

To maintain sea-level power at higher altitudes, the turbocharger must maximize engine volumetric efficiency. Volumetric efficiency is the ratio of total cylinder volume to actual displacement. For instance, if the intake stroke of a piston completely filled the cylinder with fuel/air mixture at ambient pressure, it would be 100% efficient.

Normally, aspirated engines are those that draw the fuel/air mixture into the cylinders by the partial pressure created on the downstroke of the piston. During the downstroke in a four-stroke cycle, the pressure in the cylinder becomes less than that of the ambient air, so the fuel/air mixture is "sucked" into the cylinder through an intake valve. Such engines are never 100 percent efficient, for a number of reasons.

At high RPM, the entire operation happens so quickly, the mixture simply doesn't have enough time to fill the cylinder before the intake valve slams shut. Under the best of conditions, the airflow is slowed by obstacles such as bends in the manifold. Even if 100 percent efficiency were possible, there still is the problem of decreasing air density, which accounts for as much as 50 percent loss of power at 12,000 feet.

The solution to some of those problems is the turbocharger. Instead of mechanical linkage connecting the unit to the engine, a free-turning, vaned wheel is placed directly in the engine exhaust stream, as depicted in Figure 7-2. The turbine wheel, driven by engine exhaust, is connected to the impeller portion of a centrifugal compressor assembly. The impeller directs ambient air to the spinning compressor axis. The air is spun at a very high velocity, forcing it outward and causing increased pressure. This high-pressure air then is ducted through the intake manifold of the engine to the various cylinders, as shown in Figure 7-3.

Using engine exhaust to power a turbocharger makes it seem like you're getting something for nothing, but there are disadvantages. One of the major problems is the adverse operating conditions of the turbocharger itself. The turbine and compressor routinely rotate at speeds in excess of 100,000 RPM. The turbine, which drives the entire unit, is subjected to extremely high engine-exhaust temperatures.

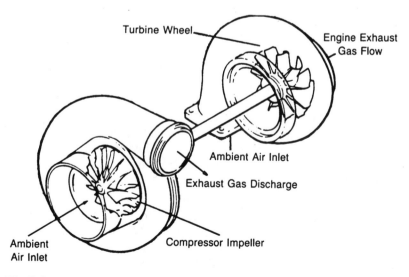

Fig. 7-2. *Basic turbocharger diagram.*

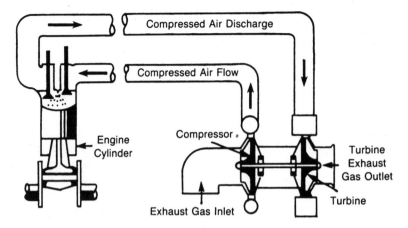

Fig. 7-3. *Turbocharger flow diagram.*

HIGH-ALTITUDE OPERATIONS

There also are special circumstances to consider when operating at the higher altitudes that become available through the use of a turbocharger. Fuel vaporization is one such problem.

Engine-driven pumps pull fuel to the intake manifold, which at high altitude invites vapor lock. So the aircraft must be equipped with tank-mounted boost pumps to feed fuel to the engine-driven pump under positive pressure. Boost-pump failure can cause cavitation, and eventually failure, of the engine-driven pump as well as vapor lock and engine fuel starvation.

Another problem associated with high-altitude operation, though not turbocharger related, is worth mentioning. It is the electrical conductivity of the rarefied atmosphere. Magnetos and wiring harnesses require special care and protection to prevent electrical problems that cause engine roughness and possible failure.

MISCONCEPTIONS

A number of misconceptions surround the use of turbochargers. It is commonly believed that turbochargers increase fuel efficiency. They don't, but they do allow you to take advantage of higher true airspeeds and more favorable winds at higher altitudes.

Some operators think the increased fuel/air mixture causes greater stress on the engine. It would seem logical, but the opposite is true; a turbocharged engine has less operating stress. Remember, a normally aspirated engine has four piston strokes: intake, compression, power, and exhaust. Compression, power, and exhaust strokes occur in a positive-pressure condition, causing fundamentally the same type of pressure on the piston, rings, and connecting rod. The intake stroke, however, creates suction, which causes a significant change of force on the piston. This force change is transferred to the crankshaft via the piston connecting rods. The faster the engine operates, the worse the effects of the pressure change.

The turbocharged engine, on the other hand, has all positive pressure strokes. When the intake valve opens, air from the turbocharger is pumped into the cylinder as fast as the piston can move downward. This prevents drastic force changes and is easier on the engine.

Despite the consistency of pressure within the cylinders, turbocharged engines have shorter TBOs than their normally aspirated counterparts. One factor used in determining TBO is how hard the engine is worked during an average hour. With a turbocharged engine, the pilot is able to operate at rated takeoff power for a significantly longer time. Whereas the normally aspirated engine begins to lose power immediately after takeoff, the turbocharged engine routinely operates at a higher percentage of its rated power during a greater part of its lifetime. That means the engine must dissipate more heat over a longer period of time. That greater amount of heat results from higher power settings, in addition to the heat generated by the turbocharger itself. Remember, when air is compressed, it increases in temperature. That hot, compressed air is mixed with fuel and shoved into the engine, and that heat reduces engine life.

CONTROLLING A TURBOCHARGER

From an operational standpoint, a turbocharger needs one additional piece of equipment. If engine exhaust gases spin the turbine, which in turn spins the compressor, the system needs a way of controlling the freewheeling turbine and the resultant air pressure. Control is accomplished with a *wastegate*, a damperlike device that regulates the amount of exhaust that hits the turbine rotor. This may be seen in the illustration of the Cessna Turbo Stationair turbocharger system in Figure 7-4.

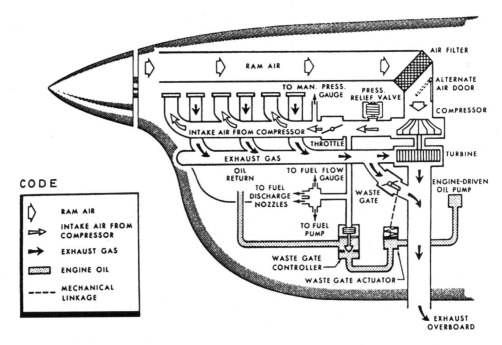

Fig. 7-4. *Cessna TU206G Turbo Stationair turbocharger system.* (Courtesy of Cessna Aircraft Company)

Chapter Seven

A full-open wastegate directs exhaust straight through the exhaust pipe and overboard, so it doesn't hit the turbine rotor and the compressor doesn't produce pressurized airflow. As the wastegate is moved toward the closed position, more and more exhaust is channeled to the rotor, which causes it to spin faster and faster. From a practical standpoint, the wastegate is like a variable slough that controls the amount of water that goes to a waterwheel. The slough can divert water away from the wheel, channel a little to it, or direct a great amount of it to the wheel.

Early turbocharger systems required the pilot to control the wastegate directly. It would be open for takeoff, and then as the engine began to lose power in the climb, the pilot would close the wastegate gradually to maintain power. Occasionally, pilots would attempt to take off with the wastegate closed. The resultant "overboost" would sometimes provide spectacular takeoff and climb performance, but like trying to stuff 10 pounds of potatoes into a 5-pound bag, the resultant excessive m.p. would eventually destroy the engine. Similarly, some pilots would forget to reopen the wastegate as the aircraft descended and an overboost would occur. Early turbocharger systems did not compensate for changes in airspeed, pressure, or temperature, resulting in disconcerting fluctuations in manifold pressure (m.p.) called "bootstrapping." As a result, three basic wastegate control systems evolved: fixed, throttle-controlled, and automatic.

Types of Wastegate Control Systems

The fixed, ground-adjustable wastegate remains in the same position throughout all engine operations. The exhaust flow is split, some of it dumping overboard and some going through the turbine. A very simple system, the pilot controls m.p. at all times with the throttle. The advantages of such a simple system are minimal maintenance and relatively low cost; however, there are some major drawbacks.

The most obvious drawback to the fixed, ground-adjustable wastegate is the significant loss of potential power in the exhaust dumped overboard. By its very nature, the system has a low critical altitude, though it is better than having no turbocharger at all. Perhaps the most important drawback of the fixed wastegate is that the throttle is the only m.p. control, therefore the compressor may produce more pressure than is necessary, thereby subjecting the engine to more heat than is good for the engine. Finally, the fixed wastegate is susceptible to the nagging problem of bootstrapping.

The next step up is a turbocharger system that has a mechanical linkage connecting the throttle to the wastegate. At low power settings, the throttle opens normally, not affecting the full-open position of the wastegate. Once the throttle reaches a full-open detent, the mechanical linkage then allows further throttle increases in conjunction with automatic closure of the wastegate. This provides some control over the turbocharger, but bootstrapping still can be a problem.

The two types of automatic wastegate controller are the density-reference system and the pressure-reference system. The pressure-reference system maintains a selected manifold pressure as set by the throttle. Engine oil pressure, directed by the controller, moves the wastegate as necessary to maintain the appropriate m.p. The obvious advantages of the pressure-reference system are that the m.p. does not need to be adjusted manually during climb, and the maximum m.p. is limited so the pilot does not have to worry about overboosting.

The density-reference system, used on some Lycoming engines, goes one step further. At full throttle, the wastegate is controlled by compressor discharge air. The pilot selects the desired m.p. and the density, or slope, controller tries to hold that density of air, regardless of changes in airspeed, ambient pressure, or temperature. This system actually can increase the m.p. by several inches during a climb, which has startled more than one unsuspecting pilot. Many pilots react to that situation by backing off on the throttle in the mistaken belief that the controller has failed and is putting out too much pressure. At less than full throttle, the turbocharger reverts to being a pressure-reference system, and things work pretty much as the pilot expects. The obvious advantage to such a system, when used correctly, is that it is automatically controlled and compensates for variation in density and airspeed. It almost totally eliminates bootstrapping.

HEAT PROBLEMS

For all the positive aspects of the turbocharger, it still has the nagging problem of hot, compressed air being pumped into the intake manifold. Not only does heat destroy engines, but it also robs them of power. A given turbocharged engine might produce the same m.p. and RPM at 13,000 feet as it does at sea level, but that doesn't mean it's producing the same horsepower. The hotter the air, the less dense it is, and while the turbocharger packs molecules of air together to overcome the increased density altitude that comes with altitude, it is not unusual for 15 to 20 percent of the engine's horsepower to be lost to heat, which means the turbocharger must work that much harder to make up the difference. This in turn creates more heat, causing a vicious circle.

To reduce the extent of the heat problem, some manufacturers put intercoolers between the compressor discharge and the intake manifold. The intercooler cools the hot, compressed air before it goes to the engine. Not only does that increase the critical altitude, but it also reduces the potential for detonation as a result of overheating. The disadvantages are increased weight, increased drag because the intercooler uses ram air, additional system complexity, and added expense.

With a turbocharger system, the manifold pressure gauge is controlled by the throttle and the tachometer is controlled by the propeller lever. The exhaust-gas temperature (EGT) gauge displays mixture control, and the cylinder head temperature (CHT) gauge is controlled by cowl flaps and airspeed. From an operational standpoint, the key to system longevity is heat control.

During takeoff and climb, EGT is controlled by throttle setting. If the EGT gets too high, power must be reduced. During cruise, EGT is a direct result of mixture manipulation; however, RPM also has an effect on EGT. If engine RPM is so fast that the exhaust valve opens before combustion is complete, EGT will be hotter because combustion will take place in the exhaust manifold.

Many pilots think takeoff and climb are the most critical periods of operation for an engine; actually, cruise can be far more critical. It is true that during takeoff, the engine is under full power, but it is also using a rich mixture, with the excess fuel acting as an additional coolant. In cruise, despite the lower power setting, a significantly leaner mixture is used, which creates greater potential for overheating. Remember that at cruise, mixture is the most important EGT control.

One of the most overlooked instruments is the cylinder head temperature gauge (CHT). During takeoff and climb, cowl flaps should be open and the proper airspeed should be flown. If the temperature becomes excessively hot with full-open cowl flaps, the pilot must increase airspeed with a shallower climb, so the engine will get a greater flow of cooling ram air. Even cruise may require partially open cowl flaps, especially when operating at economy power settings. Remember that the turbocharger is stuffing hot air into the engine.

Another confusing aspect of this situation is that cylinder cooling is reduced at high altitude. It may not seem logical because it is so cold at higher altitudes, but less dense air has very poor cooling properties. If the pilot cannot keep CHT within an acceptable range at high altitude, he or she may need to request a lower altitude to take advantage of greater air density.

Descent requires consideration, too. Turbocharged engines operate hotter at higher altitudes. This provides a tremendous potential for thermal shock during a descent. Except in an emergency, the pilot should never pull the power back to idle at high altitudes. Plan on making power descents. Use flaps and gear if necessary to increase vertical speed, but maintain power. A good rule of thumb is to use 5-inch reductions in m.p., with a couple of minutes between reductions. This allows time for the engine to adjust to the decreasing temperature.

When preflighting the engine, carefully examine the turbocharger and exhaust pipes for cracks. The engine and turbocharger should be checked for loose fittings, oil leaks, cracks, cuts, or holes. Surrounding areas should be inspected for paint blisters or corrosion, indicating excessive heat. Other potential problem areas include clogged air cleaners and clogged engine crankcase breathers. If a preheat is necessary, take sufficient time to ensure that the entire engine is heated. Often the temperature probes will be heated sufficiently to give an erroneous indication in the cockpit, while oil in the engine sump and outlying accessories still is congealed.

TROUBLESHOOTING

The single most important rule of thumb in troubleshooting a turbocharger problem is to check the engine first. Far too many units are repaired or replaced only to have the problem reappear immediately. Most turbocharger problems are caused by one of the following: lack of lubrication, foreign object damage, or contamination of lubrication. Table 7-1 provides a problem, cause, remedy troubleshooting checklist.

The most common turbocharger problem is lack of lubrication, which usually shows up first as bearing failure, wheel rub, seal damage, or shaft breakage. Foreign objects can damage either the turbine blades or the compressor, but in either case, a wheel imbalance at 100,000 RPM can be devastating. Contamination of lubricant causes scored shaft journals and bearings, blocks oil holes, plugs seals, and eventually leads to heavy oil leakage.

Table 7-1   Turbocharger Troubleshooting

Engine lacks power	Black exhaust smoke	Excessive engine oil consumption	Blue exhaust smoke	Turbocharger noisy	Cyclic sound from turbocharger	Oil leak from compressor seal	Oil leak from turbine seal	Cause	Remedy
●	●	●	●		●			Clogged air filter element.	Replace element according to engine manufacturers' recommendations.
	●	●	●	●	●	●		Obstructed air intake duct to turbo compressor.	Remove obstruction or replace damaged parts as required.
●	●		●					Obstructed air outlet duct from compressor to intake manifold.	Remove obstruction or replace damaged parts as required.
●	●		●					Obstructed intake manifold.	Refer to engine manufacturers' manual and remove obstruction.
			●					Air leak in duct from air cleaner to compressor.	Correct leak by replacing seals or tightening fasteners as required.
●	●	●	●	●				Air leak in duct from compressor to intake manifold.	Correct leak by replacing seals or tightening fasteners as required.
●	●	●	●	●				Air leak at intake manifold to engine joint.	Refer to engine manufacturers' manual and replace gaskets or tighten fasteners as required.
●	●	●	●	●	●			Obstruction in exhaust manifold.	Refer to engine manufacturers' manual and remove obstruction.
●	●		●					Obstruction in muffler or exhaust stack.	Remove obstruction or replace faulty components as required.
●	●		●		●			Gas leak in exhaust manifold to engine joint.	Refer to engine manufacturers' manual and replace gaskets or tighten fasteners as required.
●	●		●		●			Gas leak in turbine inlet to exhaust manifold joint.	Replace gasket or tighten fasteners as required.
			●					Gas leak in ducting after the turbine outlet.	Refer to engine manufacturers' manual and repair leak.
		●	●			●	●	Obstructed turbocharger oil drain line.	Remove obstruction or replace line as required.
		●	●			●	●	Obstructed engine crankcase vent.	Refer to engine manufacturers' manual, clear obstruction.
		●	●			●	●	Turbocharger center housing sludged or coked.	Change engine oil and filter, overhaul or replace turbo as required.
●	●							Fuel injection pump or fuel injectors incorrectly adjusted.	Refer to engine manufacturers' manual — replace or adjust faulty component(s) as required.
●	●							Engine camshaft timing incorrect.	Refer to engine manufacturers' manual and replace worn parts.
●	●	●	●			●	●	Worn engine piston rings or liners (blowby).	Refer to engine manufacturers' manual and repair engine as required.
●	●	●	●			●	●	Internal engine problem (valves, pistons).	Refer to engine manufacturers' manual and repair engine as required.
●	●	●	●	●	●	●	●	Dirt caked on compressor wheel and/or diffuser vanes.	Clean using noncaustic cleaner and soft brush. Find and correct source of unfiltered air and change engine oil and oil filter.
●	●	●	●	●		●	●	Damaged turbocharger.	Analyze failed turbocharger, find and correct cause of failure, overhaul or replace turbocharger as required.

Chapter Seven

There is little troubleshooting that can take place in flight. The system either works or it doesn't. If there is a sudden loss of m.p., the turbocharger is the likely culprit, and you will have just reverted to a normally aspirated engine. That alone is not a significant problem, unless you are over high mountains. The most important question is, why did the turbocharger fail? Watch for fire!

In the event of a turbocharger failure, carefully monitor pressures and temperatures and watch for a loss of engine oil. Any sign of either fire or loss of oil merits an immediate engine shutdown. One other preventive measure you can take is to listen to the system. Become familiar with the sound of your turbocharger; it has a high whistle or whine that can be very eerie sounding when flying at night in the clouds; come to think of it, everything is eerie sounding when flying at night in the clouds. Anyway, should that sound become abnormally loud or shrill, there probably is insufficient bearing clearance and the unit needs servicing to head off a failure.

The best preventive maintenance is to catch problems before they become major. At 100,000 RPM, little problems become big problems fast. Oil leaks, unusual vibrations, and sounds should be followed up carefully. Nothing is as effective as good operating procedure. When shutting down the engine, the oil pressure drops to zero, and so does the lubricating ability of the turbocharger. A typical approach and landing is at low m.p., allowing plenty of time for the unit to spool down. But, if high m.p. is carried on the approach, or used during taxi, the turbine still will be spinning after engine shutdown, but there will be no lubrication for it. The prudent pilot will set the engine at idle for several minutes prior to shutdown to assure adequate lubrication and spool-down time.

It is also important to change oil at least at the frequency prescribed in the pilot's operating handbook (POH). The bearings are highly susceptible to dirty or improper oil. Turbochargers should be overhauled at the same time the engine is, at the recommended time between overhaul (TBO). In the long run, this reduces unscheduled downtime and gives the owner a lower operating cost.

Spinning turbocharger parts are machined to within one-millionth of an inch. It is a good idea to conduct any required maintenance or inspections on time. Assure that there is no buildup of exhaust carbon or warpage in the wastegate. Pilots who operate in dusty or smoky air should be especially careful to change the oil frequently and follow prescribed inspections. If it sounds as if turbochargers are problematic, then you should realize that at least one manufacturer has no scheduled periodic maintenance requirement for its turbochargers. If you follow normal engine maintenance and keep the oil clean, there should be no problems at all!

Finally, most modern turbocharged engines have overboost protection. Don't rely on it! Always allow a minimum of 30 seconds warm-up before running up an engine, longer when it's cold; this prevents oil lag. On takeoff, advance the throttle gently to about 25 inches while holding the brakes. Let the turbocharger come up to speed, then release the brakes and gently advance the throttle to the m.p. indicated in the POH for takeoff.

8
Electrical Systems

ELECTRICAL SYSTEMS HAVE BEEN AN ONBOARD FIXTURE OF AIR-
craft since the Wright Flyer. In those days the role of electricity was limited to the
magneto providing sufficient voltage to spark the fuel/air mixture. As demand increased for electricity to power other equipment, first the wind-driven generator was
developed and subsequently its engine-driven version.

For many years, light planes exclusively used the 14-volt electrical system. Starting approximately in the early 1980s, the 28-volt system began to take over light aircraft. In either case, the fundamentals are the same. The primary purpose of igniting
the fuel/air mixture is still the exclusive, and independent, domain of the magneto.
However, the demand for electrical energy in the airplane has increased tremendously.

Even the simplest electrical systems in modern light aircraft power engine starters,
cockpit and position lighting, navigation and communication equipment, engine and
flight instruments, and accessories ranging from cigarette lighters to in-flight tele-
phones. Unfortunately, as the complexity of the electrical system increased over the
years, the average pilot's understanding of it decreased. While we can leave the theory
of electrons, neutrons, protons, and so-ons to the engineers, a basic understanding of
the fundamentals of electrical systems is necessary if a pilot hopes to be able to ade-
quately preflight or determine in-flight problems.

ELECTRICAL SYSTEM THEORY

If you are reading this chapter with the hopes of answering the question "What is electricity?" you are going to be disappointed. In many ways, electricity is still a mystery. Our knowledge has significantly increased about the subject since we discovered that lightning is electricity, and at least most of us no longer believe lightning is a sign of the displeasure of some god. But electricity, like lightning, still evokes uneasiness in most and fear in many.

Electrical Circuits

Perhaps what makes electricity so mysterious is that it is intangible. We can't hold it in our hand or, for the most part, even see it. Electricity is a name for a collection of laws found in a physics book. But we do know about how it behaves and how we can get it to perform ever increasingly complex tasks. For the pilot, the important issue is not so much how electricity works but rather how electrical circuits manage to do the kinds of things they do in an aircraft.

Series Circuits

Perhaps the easiest way to understand circuitry is to think of it as rivers of electricity. Figure 8-1 illustrates the three basic types of circuits: series, parallel, and series-parallel. In the series circuit, current flows from the primary bus to the 5-amp circuit breaker (CB), then to the fuel quantity indicator. Therefore CBs are always placed upstream, and in series with, the circuit or accessory they are to protect. If the CB fails, all electric current is cut off from anything "downstream." It was precisely this type of system that evoked the cheers and jeers of my sisters and me each Christmas as we would try to find the burned-out bulb in our ancient Christmas tree lights. One burned-out bulb in a series circuit would prevent the whole string from lighting and officially beginning the Lombardo family Christmas season. Those under thirty may not be able to relate to that time-honored tradition because modern tree lights are parallel-wired.

Parallel Circuits

In parallel circuits, current flows from the primary bus simultaneously to both 5-amp CBs and then on to both the landing light and fuel quantity indicator. Failure of one circuit, say the landing light, will in no way affect the fuel quantity indicator circuit. This is the ideal situation in an aircraft, and certainly all critical electrically driven components operate in parallel with other components. Unfortunately, there are many electrical circuits in modern aircraft, and individual protection of each item is not always practical or necessary.

Series-Parallel Circuits

In the series-parallel circuit, a number of items may be put in parallel with each other, such as the taxi and landing lights, but in series with a single CB. If one of the lights burns out, the other will continue to work. Should a short circuit somehow develop in one of the lights, which could endanger the electrical system, both lights will be isolated from the system when the CB fails.

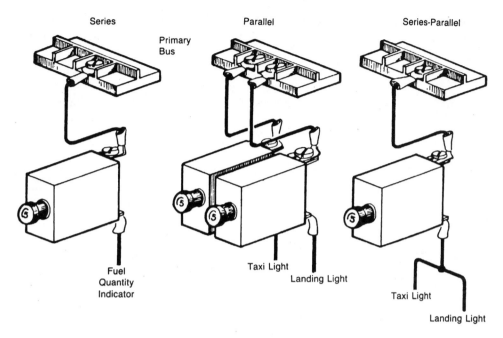

Fig. 8-1. *Electrical circuits showing primary bus and circuit breakers.*

SOURCES OF ELECTRICAL POWER

Aircraft have very limited choices with respect to a ready supply of electricity. Light, general-aviation aircraft do not have the luxury of onboard auxiliary power units like most turbine engine aircraft. In the early days of aviation, aircraft did not have storage batteries to run a starter, so the pilot had to turn the propeller by hand to get the engine to start. Then, as now, the independent magneto system was geared to the engine, so "propping" the engine would cause the magneto to turn, with a resulting timed spark to the cylinders. Today's high-compression engines make propping next to impossible and certainly an unsafe practice.

We now have three possible sources of electricity available to light aircraft. At most airports we have access to a ground power unit (GPU). Onboard the aircraft, we have a storage battery and alternator.

THE GROUND POWER UNIT

The ground-based source of power to an aircraft is the ground power unit (GPU). It is often incorrectly referred to as an APU, but the difference is not academic. The GPU is exactly what its name implies—a ground power unit. It is usually mounted on wheels and is often operated by a flight line attendant. The APU (auxiliary power unit) is an additional small engine, usually a turbojet whose sole purpose is to supply auxiliary power (electric and often pneumatic) to run aircraft systems on the ground when the engines are shut down and to aid engine start-up. APUs, which are permanently installed onboard, are found on large aircraft.

Today, virtually all GPUs have the standard NATO 3-pin plug that neatly inserts in a "conveniently" located receptacle on the airplane, all too often being directly behind a propeller or under the wing near the fuselage.

Ground power is typically used to start an engine for the first time on a cold day— conditions difficult for the battery to operate in. The GPU is also used by mechanics when they need electrical power for extended periods of time without the engine running. When using a GPU there are several areas of concern. First make sure all avionics are turned off.

Avionics are highly susceptible to damage from transient high voltage, or "spikes." Most GPUs have a variable voltage output so it is necessary to make sure the proper voltage has been selected on the power unit to match your aircraft electrical system. Similarly, the polarity of the GPU must match your aircraft system's polarity. Some aircraft incorporate polarity reversal protection in the aircraft receptacle, however the prudent pilot will always check it before utilizing the power source. For reasons that will become apparent later, GPUs should not be used to start an aircraft that has a dead battery.

THE LEAD ACID STORAGE BATTERY

The heart of the electrical system is the battery, and lead acid batteries are still standard equipment in light, general-aviation aircraft.

It's important to understand that the battery is not an electricity producer; it is a storer. The modern aircraft lead acid battery is an efficient, carefully designed piece of equipment that is significantly different from its automotive cousin. It must operate dependably under conditions unheard of for auto batteries. Its list of jobs includes engine starting, preflight of electrical equipment and accessories, and backup for the alternator. The conditions under which it must operate are extreme, to say the least. Aircraft go where even cars dare not tread: the arctic tundra, high up in the mountains, far into deserts, deep into rain forests, onto rivers and lakes, and into airports below sea level.

The battery must operate reliably in unusual attitudes, including inverted flight, at very high altitudes, and be capable of handling potential temperature changes in excess of 100 degrees during a single flight. Despite all of its capability, the aircraft battery, as seen in Figure 8-2, weighs less and is smaller in size than its automotive counterpart. There are trade-offs, though, and the aircraft battery tends to be a little more delicate and does not maintain itself as long. The aircraft battery is designed to provide a greater, short-term capacity such as the cold weather start. To understand what a battery can do for you, and what you must do for it, some theory is necessary.

Theory of Operation

Three important terms used when discussing the battery are volts, amps, and amp-hour. A *volt* is a measure of electrical pressure; it is potential. It is the motivating force that moves electrons through a conductor. A 12-volt battery has a potential of 12 volts of electrical pressure. *Amps* describe flow rate, or current—a measure of how many electrons flow through a conductor. *Amp-hour* is a rating given to a battery indicating potential

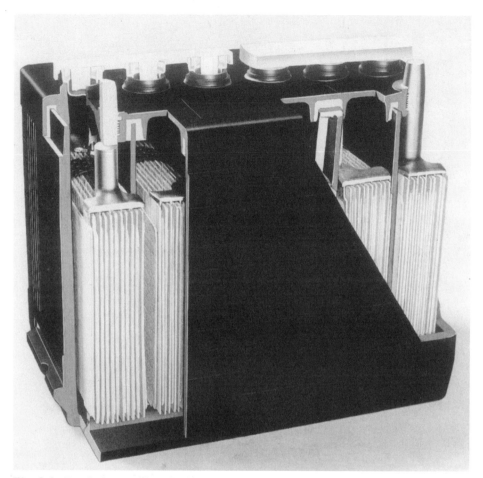

Fig. 8-2. *Lead plates within the battery provide a high surface-area-to-volume ratio for compact size and high efficiency.*

duration of the current flow under ideal conditions. You may think of the whole process as a miniature waterwheel like the one shown in Figure 8-3.

The upper water source and lower water receiver are like the terminals of a battery. As the water flows from one to the other, it expends energy, turning the waterwheel, or in the case of a battery, turning a flap motor or illuminating a landing light. The greater the load on the waterwheel, the greater the demand for flow to turn it. Similarly, the more powerful the electric motor, the greater the demand for current. Eventually, you will exhaust the supply of water in the tank, and so too will you exhaust the battery.

A fully charged, 20-amp/hour battery in ideal conditions, for example, would be capable of supplying 20 amps for 1 hour, 10 amps for 2 hours, or 5 amps for 4 hours. In our water system we have a pump that will resupply the source. In the aircraft we have an engine-driven alternator that will recharge the battery continuously, whenever the engine is

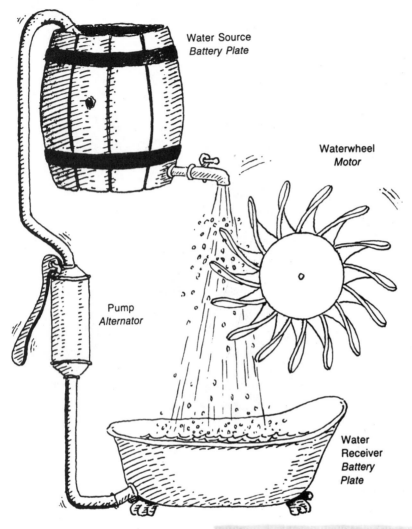

Fig. 8-3. *Analogy of a waterwheel to a lead-acid battery.*

running. The alternator, by regulation, is also sufficient to handle normal in-flight power requirements. If the alternator system fails, then we have a limited supply of energy available from the battery.

Murphy's Second Law states, "Nothing is ever as simple as it first seems," and batteries neatly fall under that law. It is a very common misconception that a battery stores electricity; it's easiest to think of it in that manner. But a battery actually converts electricity into chemical energy, stores the chemistry, and reconverts it when the battery is connected to some demand such as a hungry starter. Understanding the basic chemical process gives a clue to potential problems. The four primary chemicals involved are lead dioxide in the positive plates, a spongy form of lead in the negative plates, and a liquid electrolyte composed of sulfuric acid and water.

When the battery terminals are connected externally to an electrical demand, the sulfate in the electrolyte combines with the active material in both types of plates, and the electrons will flow from the negative terminal through the starter to the positive terminal. At the same time, both plates will be accumulating a coating of lead sulfate, which is highly resistant to current flow. As the coating increases, the battery gets weaker until both plates are completely coated, then chemical action ceases and the battery stops working. The phenomena that makes the battery practical is that by reversing the current flow, the entire process works in reverse and recharges the battery, which is one of the jobs performed by the alternator in flight.

Lead Acid Battery Characteristics

The battery, like the human body, reacts differently to varying environmental conditions. During cold weather it does not produce as much voltage because the chemical reaction slows down. Referring to Table 8-1, you will see that it is not uncommon for a lead acid battery to lose as much as 40 percent of its warm weather capacity during winter months in the Snow Belt area. It is also important during winter to change to a lighter grade of oil and to change it often to keep it clean. The colder it gets, the stiffer the oil becomes, and the engine with cold, dirty, heavy-grade oil will be very difficult for the battery to start. Pulling the prop through several times helps break up the oil, but the best course of action is to warm the engine oil with a preheat and perhaps even use a ground power cart for starting.

To determine the state of charge of a battery, it would seem logical to simply put a voltmeter on the battery's terminals and read the open circuit voltage, but remember Murphy's Second Law! Temperature has a very strong effect on cell voltage and there is no convenient way of correcting for it. Determining the specific gravity (SPGR) of the electrolyte is a more reliable method. It is true that temperature also affects SPGR; however it is easily calculated by referring to Table 8-2, which shows that the hotter the electrolyte, the lower its SPGR.

Table 8-1.
Percent of Battery Power Available
at Varying Temperatures

Temperature °F	Full Charge	Half Charge
80	100%	46%
32	65%	32%
0	40%	21%

Table 8-2.
How Temperature Affects Specific Gravity

°F	Specific Gravity
107	1.260–1.280
77	1.280–1.290
47	1.290–1.300

To test the SPGR of the electrolyte, it will be necessary to use the fairly inexpensive and readily obtainable hydrometer shown in Figure 8-4. As the battery produces current, the acid transfers from the electrolyte to the active material in the plates, therefore less acid remains in the electrolyte. A chemist will tell you that the specific gravity of acid is considerably greater than the SPGR of water, so the loss of acid causes a drop in the specific gravity of the electrolyte.

To discover the SPGR of the electrolyte, you will first need to take off the battery box cover as shown in Figure 8-5. With the battery exposed, take a moment to check its general condition, as in Figure 8-6. Then take the cap off of a battery cell and draw electrolyte up into the hydrometer as illustrated in Figure 8-7. The small numbered stem will float inside the instrument and the SPGR reading is taken at the fluid level. It is important to note that the stem must be floating for the reading to be accurate, therefore it must have liquid below and around it. A fully charged cell should read between 1.275 and 1.300, depending on the manufacturer, and it is a good idea to check all the cells of the battery. Readings from 1.200 to 1.240 indicate a low state of charge, and engine starting may be difficult if not impossible.

If the temperature of the electrolyte is between 70 and 90 degrees F, the reading should be accurate. For temperatures above or below that range, a correction factor must be applied, as set forth in Table 8-3. For instance, if the electrolyte temperature is 100 degrees and the specific gravity reading is 1.267, the chart tells us that there is a +.008 correction factor. 1.267 + .008 = 1.275 SPGR.

Lombardo's First Law of Reciprocal Reliability states, "The reliability of any person or thing is directly proportional to the time available to accomplish the task." Practically speaking, that means batteries tend to fail most often when you are in a hurry. Therefore, the temptation is great to put the battery on a fast charge. Be careful; there is a very strong possibility of overheating the battery and buckling the plates. Unless you are an experienced mechanic, you should limit the charging rate to 4 amps per hour. It is also worth noting that battery charging produces the highly explosive hydrogen gas. Be very careful when connecting and disconnecting the charger to prevent possible sparking, and, of course, smoking should never be permitted in the area! One word of caution. When connecting a battery to a charger, be very careful about polarity. Reversing polarity will destroy a battery. Note that in Figure 8-8 a cautious mechanic went so far as to label the cables with "positive" and "negative."

My best friend and male golden retriever Roscoe used to have a thing about my leaving him at home alone. He would go into my bathroom and carefully remove everything from my wastebasket and lay it on the floor. Batteries are a lot like Roscoe used to be; if left alone too long, they tend to cause problems. A battery not used for a month can easily self-discharge to 50 percent capacity. Temperature is a significant contributing factor to the rate of self-discharge; the higher the temperature, the faster the specific gravity decreases—as much as .003 per day with an ambient temperature of 100 degrees F.

- 100 degrees F— .003 SPGR per day.
- 80 degrees F— .002 SPGR per day.
- 50 degrees F— .0005 SPGR per day.

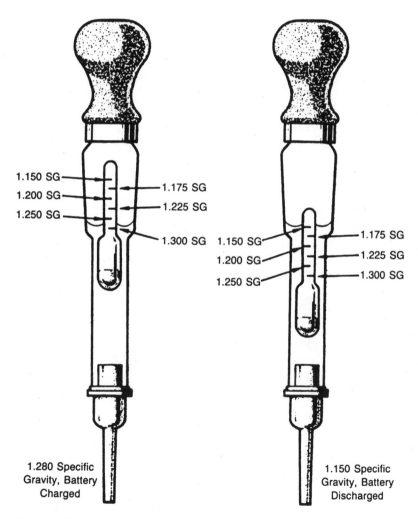

Fig. 8-4. *Typical hydrometer readings.*

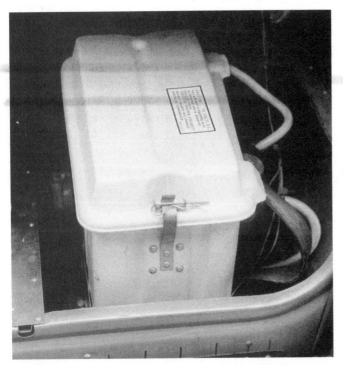

Fig. 8-5. *Remove protective covering to access the battery.*

Fig. 8-6. *Remove cover and check cable contacts for corrosion and security.*

Fig. 8-7. *Testing the electrolyte's specific gravity with a hydrometer.*

From the list above it would appear the worst time to let a battery sit idle is the hot months. However, here again, Murphy's Second Law rears its ugly head. The chemical process involved performs a nasty little trick; it causes oxygen to enter the electrolyte and mix with the hydrogen already present. A quick trip to the old high school chemistry book assured me that water is the result. So, in winter, a discharging battery keeps adding pure water to the electrolyte, which, when exposed to freezing temperatures, means a cracked battery.

A fully charged battery with an SPGR of 1.285 would freeze at –90 degrees F, but a look at Table 8-4 reveals that a battery with a 1.100 SPGR has a freezing point of only 19 degrees F. The lesson to be learned is whenever you anticipate not flying regularly, you should remove the battery and store it in a cool place. In addition, approximately every five weeks it should be recharged to prevent a lead sulfate buildup on the plates. Lead sulfate (a crystalline formation) is a terrible conductor, which may prevent recharging and permanently damage the battery.

Battery Installation

When installing a new battery there are several important considerations, the most important of which is reading the manual supplied by the manufacturer. In all cases, follow the manual! Only a fully charged battery should be put into an airplane, and few, if any, batteries come fully charged. According to the Gill GSM-682 Service Manual, new batteries are received dry-charged and will deliver 75 percent of the rated capacity after the

Table 8-3.
Temperature Correction for
Specific Gravity Readings

Electrolyte Temp. (°F)	Correction Factor
120	+.016
110	+.012
100	+.008
90	0
80	0
70	0
60	−.008
50	−.012
40	−.016
30	−.020
20	−.024
10	−.028
0	−.032
−10	−.036
−20	−.040
−30	−.044

Fig. 8-8. *Note polarity marked on each charger cable.* (Photo by author, courtesy of Frasca Air Services)

Table 8-4.
Effect of Specific Gravity on Battery Freezing Temperature

Specific Gravity Corrected to 80°F	Battery Freeze Point (°F)
1.289	−90
1.250	−62
1.200	−16
1.150	5
1.100	19

initial filling of electrolyte without further charging. However, it is important that they be given an initial charge to ensure their airworthiness before installation in the aircraft. The reason that they should be charged on a battery charger before being installed in the aircraft is that the alternator will charge the battery at a very high rate. Charging from so low a state of charge at such a high rate leads to battery overheating and potential buckling of cell plates. Another significant concern when installing a new battery is the mixing of the electrolyte.

I remember well a cute trick I played on my high school chemistry lab partner. I threw a vial of water into a beaker of acid—lots of fun for everyone except the girl downstream and her new angora sweater. She received both a second-degree burn and a sweater modified by Lombardo, and I received an "F" in the course for that and other various and sundry reasons. Moral of the story: Always pour acid into water, not water into acid! And remember, it really is sulfuric acid. If you spill a little on the airplane you will discover that it literally eats through aluminum at an alarming rate. If that doesn't bother you, consider that it will also cause severe skin burns and blindness, and inhaling the fumes may cause you to permanently lose your sense of smell and taste.

If acid comes in contact with your skin, immediately wash the affected area with baking soda to neutralize it. Needless to say, sulfuric acid is very powerful, so much so that the simple act of mixing it with water into electrolyte creates heat. Always give the mixture time to cool off before actually pouring it into the battery cells to prevent potential plate damage from overheating.

As a battery charges and discharges over time, the active material from the plates wears off and slowly collects at the bottom of the battery. Like people, batteries get old too and at some point the plates will no longer have enough active material left to produce adequate capacity. When you begin to find that the battery just doesn't hold a charge like it used to, then it is probably time to trade it in for a new one. Due to the tendency for active material to wear off and collect at the bottom, it is important never to attempt to remove the electrolyte by turning the battery upside down and letting it pour out. The sediment will then run over the cell plates, creating a strong potential for the plates to short out. If for any reason you need to remove the electrolyte, it should be drawn out with a syringe-type instrument or even a hydrometer.

For all its good points, the lead acid battery still has many limitations and the nickel cadmium battery (NiCd) is finding its way into more and more aircraft.

THE NICD BATTERY

The nickel cadmium (NiCd) battery, referred to as "nicad" (pronounced "nye-kad"), works under the same principle as its lead-acid counterpart. Both are electrochemical systems that store and supply electrical energy as needed.

Designed for a long life under extremely adverse conditions, the NiCd has a very low freezing temperature, minimal gas emission during operation, and, unlike the lead acid battery, suffers no deterioration if left in a discharged state. Unfortunately, its high initial cost has hindered its growth into light aircraft.

The main components of the NiCd battery are positive and negative plates, separators, the electrolyte, the cell container, and the cell vent. The individual cells—typically 19 or 20, depending on application and manufacturer—contain thin, porous, sintered nickel plates. Polarity is achieved by impregnating the plates with nickel-hydroxide for positive polarity and cadmium-hydroxide for negative. Opposing plates are separated from one another by a continuous strip of porous plastic that serves as a reservoir for the electrolyte and a sheet of cellophane, which acts as an insulator. This sandwich structure is fitted into a case, which typically is made of a plasticlike substance called *plyamide*. The negative plates are then connected to a negative cell terminal and the positive to the positive cell terminal. The two terminal openings are sealed to the case with an external O-ring to prevent electrolyte from leaking out of the case. The electrolyte is a 30 percent solution (by weight) of potassium hydroxide (KOH) in distilled water.

Each cell has a removable vent cap, which is used to service the electrolyte, as depicted in Figure 8-9. The cap serves as a one-way valve to vent gas resulting from accidental overcharging, yet prevents air from entering the cell. Individual cells are connected in series along highly conductive bars, often made of nickel-plated copper. The current can flow in only one direction.

Because each cell averages 1.2 volts, a 20-cell battery would equal 24 volts. The entire assembly, as shown in Figure 8-10 and Table 8-5, is put into a battery case, which typically is made of stainless steel, plastic-coated sheet steel, or painted sheet steel. The case is attached carefully to a ventilation system to prevent gas buildup as a result of inadvertent overcharging and to permit forced-air cooling during operation. The two end cells are connected, usually by nickel-plated copper links, to the positive and negative battery terminals.

During the discharge process, some electrolyte is drawn into the plate sets, causing the fluid level to decrease. Recharging the battery forces the electrolyte back out of the plates, causing the fluid level to rise to its normal level. The pilot should NOT add electrolyte to an NiCd battery just because the fluid level appears low. Recharging during subsequent operation could force the highly caustic electrolyte to overflow and damage the battery and surrounding area. And if the vent cap happens to be blocked when that occurs, the result could be a cell explosion.

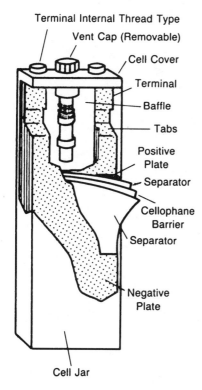

Fig. 8-9. *NiCd vented cell.*

Battery Rating System

NiCd batteries have a conventional rating system. First the nominal discharge voltage is given, 1.2 volts per cell, and then the battery capacity. The latter is the amount of power a fully charged battery provides until a fully discharged condition occurs and is measured in amps per hour (Ah). With the NiCd battery, however, it is important to know the duration, in hours, of the discharge. A nickel-cadmium battery is considered discharged when cell voltage decreases to 1.0 volts.

For instance, a 40-Ah battery at five-hour rate implies the battery would provide a total of 40 amps over a discharge period of five hours, that is 8 amps per hour. The catch is it doesn't mean the battery will provide 40 amps for one hour or 20 amps for two hours. The Saft operating and maintenance manual says the higher the current drain, the less the battery capacity. The same battery at the one-hour rate would have only a 34-Ah capacity, meaning it would supply a 34-amp current flow for one hour before the cell average dropped to 1.0 volts.

Under equal conditions, NiCd batteries long outlive lead-acid batteries. They provide faster starts, recharge more quickly, and maintain a higher state of charge longer. While the optimum temperature range is listed at 60 to 90 degrees F, with recharge cautions below –20 degrees and above +120 degrees F, NiCd batteries operate more efficiently at

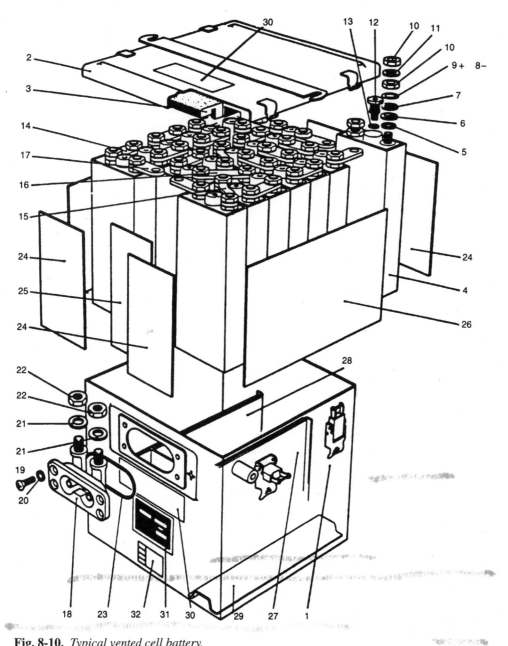

Fig. 8-10. *Typical vented cell battery.*

Table 8-5. Key to Figure 8-10

Figure No.	Nomenclature	Quantity Per Battery
1	• Complete container	1
2	• Cover	1
3	• Cover Gasket	1
4	• Cell	20
5	O-Ring	40
6	Washer	40
7	Spring Washer	40
8	Negative Polarity Washer	20
9	Positive Polarity Washer	20
10	Nut	80
11	Spring Washer	40
12	Vent Plug	20
13	O-Ring	20
14	• Connector	15
15	• Connector	3
16	• Connector	1
17	• Connector	2
18	• Battery Socket ELCON Type BR 8-1	1
19	Screw	4
20	Concave Lock Washer	4
21	Spring Washer	2
22	Nut	2
23	• O-Ring	1
24	• End Wedge	5
25	• End Wedge	1
26	• Side Wedge	2
27	• Insulating Spacer	2
28	• Insulating Spacer	2
29	• Insulating Strut	3
30	• Instruction Plate	2
31	• Identification Plate	1
32	• Modification Record Plate	1

temperature extremes than do lead-acid batteries. Many operators use NiCd batteries routinely at temperatures ranging from −30 through 130 degrees F. Other advantages are reduced engine wear as a result of faster starts, low internal battery resistance, and an inherent ability to maintain high-power output longer than an equivalent lead-acid battery.

If all this sounds too ideal, understand that cost alone is a significant deterrent to owning an NiCd battery. NiCds are often four to five times as expensive as a lead-acid battery. In addition, servicing is more critical and the battery requires constant temperature monitoring when in use. The worst drawback, though not a common problem, is called thermal runaway.

Thermal runaway

More properly called overcharge runaway, thermal runaway results in the self-destruction of the battery. The causative factors are heat, reduced resistance and current flow. An overcharge runaway scenario goes something like this. The aircraft has been flying short trips all day in instrument conditions. There have been frequent battery-powered engine starts and the electrical load was continuously heavy during start and in flight. The battery has been getting a heck of a workout. This causes a constant, excessive charging of the battery, which generates excessive battery temperature. These factors, combined with as few as one bad plate in a cell, give you the beginning of the NiCd battery's equivalent of core meltdown.

When plates short, they overheat, causing the entire cell to overheat. It, in turn, overheats the surrounding cells and as internal battery heat increases, it eventually will begin to decrease the cell's internal resistance. Lower resistance permits higher current flow. The more current that flows, the more heat produced and the lower the resistance. Before long, the cellophane between the plates in the bad cell deteriorates, making the situation worse. With even less resistance, more current flows and this begins a vicious cycle of increasing temperature and decreasing resistance. Thus far, it is the generator that is feeding the battery. If the problem is caught in time, you merely isolate the battery from the generator and the problem should end. But there is a possible outcome that is significantly more serious.

At some point, the resistance of the bad cell will be low enough that the good cells will have sufficient power to feed it. When that happens, isolating the battery does not remedy the situation because the battery is feeding itself; it is literally self-destructing! At this point the pilot can only hope to land the airplane as quickly as possible. Such a condition has been known to cause a fire in the battery box and, in at least one case, the battery actually melted through the box and dropped out of the bottom of the aircraft. Fortunately it is not a common event, and good operational and preventive maintenance procedures are strong deterrents. To aid the pilot in heading off such a problem, a monitoring system is installed with NiCd battery systems. Depending on the manufacturer, it measures either rate of battery charge (a good indication of battery heat buildup) or actual battery temperature.

THE ALTERNATOR

As the demand for a reliable electrical source increased, the wind-driven generator gave way to an engine-driven version. Though it solved all the pressing problems of the time, it too was marked for obscurity. There are several reasons why virtually all modern aircraft come equipped with alternators rather than the once-popular DC (direct current) generators.

Generators are large, heavy pieces of equipment in contrast to an alternator, which has a higher power-to-weight ratio. Another concern about the generator was the output variability with engine speed. An alternator produces more power at a lower engine RPM than a generator does—even at ground idle—a significant benefit for an aircraft sitting at the end of a runway in IFR conditions with a panel full of electronic equipment. This is accomplished by the voltage regulator, which monitors system voltage and makes the necessary adjustments to the alternator. Set correctly in a 14-volt system, for instance, the alternator will maintain system voltage +/– .5 volt.

It may help you to think of the voltage regulator as the electrical system's answer to a constant-speed prop governor. Electrical system current is fed to the alternator's exciter field in varying strengths. The more current applied, the greater the AC output. The one drawback to the alternator is that there must be amperage before it can produce electricity.

Unlike DC shunt generators, which contain a permanent magnet and have residual magnetism, the alternator rotor has little or no residual magnetism to form the necessary lines of flux to start current flow. The alternator rotor requires approximately 2 amps to make the alternator self-excite and produce current. Think of it as something like using water to prime a dry pump. Before starting the engine, this "priming" voltage is supplied by the battery to the alternator field. You can trace this in Figure 8-11, which shows that activation of the battery master switch also activates the alternator field switch. Current from the battery goes to the bus, then flows through the alternator field switch to the rotor field of the alternator. Once the engine is running and the alternator is producing current, the stator output flows to the bus and, because it is a higher voltage than the battery,

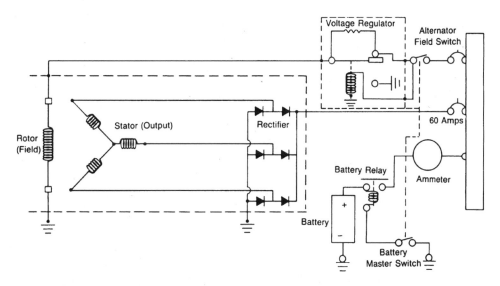

Fig. 8-11. *Light aircraft alternator system.*

the current flows into the battery, recharging it while simultaneously supplying its own field rotor with the required 2-amp excitation.

Most aircraft systems use DC voltage, so the alternator's AC output is converted (rectified) to DC through an integral silicon diode rectifier. The diodes act as one-way doors with very high resistance to current flow in one direction and low resistance in the other. They only permit current flow from the alternator to the battery. The diode prevents AC flow reversal, thereby rectifying it to DC. The DC voltage resupplies the battery and is directed to the aircraft's main bus through a 60-amp circuit breaker, where it supplies the electrical system. It also serves as the rotor's exciter current.

As the demand on the alternator increases or decreases, voltage is varied by a regulating field current. DC output is fed to a regulator-voltage-sensing coil via the 2-amp circuit-breaker alternator field switch. The coil works like a governor, sensing electric bus voltage and varying system resistance as necessary. If the bus voltage is too high, the sensing coil shifts the position of a movable contact, which puts a resistor in the circuit. This reduces the field excitation, thereby reducing the alternator output voltage. If the bus voltage is too low, the sensing-coil spring pulls the contacts back and removes the resistor from the circuit, increasing field excitation, thereby increasing alternator output voltage.

Newer alternators use solid-state regulators. This type of regulator replaces the sensing-coil mechanism and moving resistor contacts with transistors and a zener diode. The transistor acts as an electric current on/off switch and has no moving parts. The zener diode allows current to move in only one direction, except at a specified voltage value, then it reverses direction. This serves as a voltage-sensing device to vary current in conjunction with the transistors.

The modern light aircraft alternator is almost identical to its automotive counterpart. The only real difference is the aircraft alternator has a holder and special brushes for high-altitude operation. Figure 8-12 shows that the rotor is a single-field coil encased between a pair of four-poled iron sections on a shaft with insulated slip rings at one end and a nut and washer at the other. This is connected to a drive pulley that runs off the engine accessory section (or crankshaft).

Typically, the alternator field switch is interlocked with the battery master switch, which allows the pilot to shut down the alternator without turning off the battery. When the battery master is turned on, the bus bar is energized. This sends power through the alternator field switch to the brushes and slip rings. The primary difference between the brushes on an alternator and those on a generator is the alternator brushes handle only a few amps, whereas the generator brushes handle up to 60 amps. From the slip rings the power goes to the rotor. The rotor turns inside the stator, and its magnetic field cuts through the stator's three, single-phase windings, which are spaced so that the voltage induced in each is 120 degrees out of phase with the voltages in the other two windings. This Y-shaped arrangement is what produces the three-phase AC, and system voltage builds up rapidly. From there the power goes to ground, completing the circuit.

Today, because of the requirements of FAR 23, it is common for light, single-engine aircraft to come standard with a 60-amp alternator to meet the heavy demand for in-flight power. According to FAR Part 23, Airworthiness Standards—Normal, Utility,

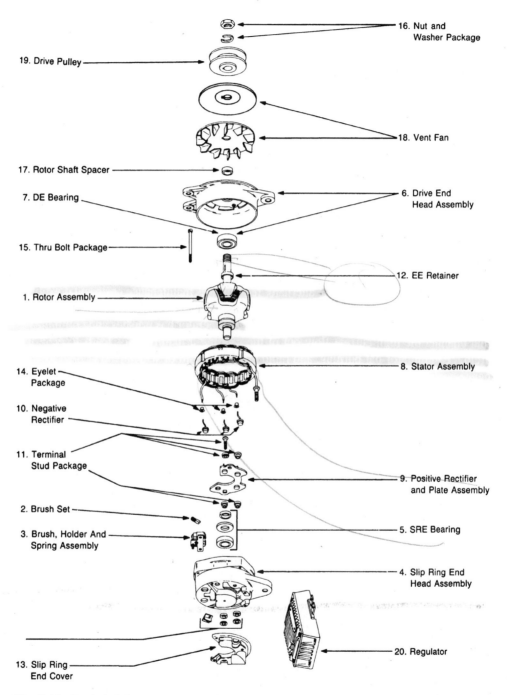

16. Nut and
    Washer Package

19. Drive Pulley

18. Vent Fan

17. Rotor Shaft Spacer

7. DE Bearing

6. Drive End
   Head Assembly

15. Thru Bolt Package

12. EE Retainer

1. Rotor Assembly

8. Stator Assembly

14. Eyelet
    Package

10. Negative
    Rectifier

11. Terminal
    Stud Package

9. Positive Rectifier
   and Plate Assembly

2. Brush Set

3. Brush, Holder And
   Spring Assembly

5. SRE Bearing

4. Slip Ring End
   Head Assembly

20. Regulator

13. Slip Ring
    End Cover

Fig. 8-12. *Exploded diagram of an aircraft alternator.*

and Aerobatic Category Airplanes, the regulation under which most general-aviation light single-engine aircraft are constructed, the following requirements exist:

23.1351 General

(a) Electrical system capacity. Each electrical system must be adequate for the intended use. In addition—

(1) Electric power sources, their transmission cables, and their associated control and protective devices, must be able to furnish the required power at the proper voltage to each load circuit essential for safe operation; and

(2) Compliance with (a)(1) of this section must be shown as follows—

(i) For normal, utility, and acrobatic category airplanes, by an electrical load analysis or by electrical measurements that account for the electrical loads applied to the electrical system in probable combinations and for probable durations...

The electrical system has to be capable of handling anticipated electrical requirements plus be able to keep the battery in a state of constant charge. Another advantage of the AC alternator was its 3-phase characteristic permitting electric motors to shed precious pounds without losing power. Finally, the alternator requires less maintenance costs and downtime.

High on the list of reasons for reduced maintenance is the fundamental design difference between alternators and generators. A generator has brushes and commutators that channel the high-current flow out of the rotating armature. This leads to electrical arcing, commutator-bar burning, and rapid brush wear, all of which lead to high maintenance costs. Also, the rotating part of the generator is heavy, which leads to greater wear on the bearings.

The alternator does not have these problems because it is connected to the external circuit by slip rings instead of a commutator, and the armature (stator) is a stationary member. The electromagnetic field (EMF) becomes the rotating member (rotor) and turns within the stator. In an alternator, the high current of the stator can go through a set of fixed leads rather than through brushes and a rotating commutator, as it does in a generator.

In all fairness, the alternator has some problems, too. Alternating current (AC), because of its expanding and collapsing nature, causes "noise" in avionics. The solid-state alternator is, in general, more prone to electrical damage than the generator, which has mechanical relays. Improper polarity can literally destroy an alternator. And an otherwise healthy regulator can burn out as a result of an unrelated alternator problem.

From an operational standpoint, there is one significant drawback to the alternator: it requires approximately 2 amps of electricity provided by the battery for it to work. Once the alternator begins to produce current, it becomes self-exciting and will continue to run, despite the condition of the battery. But if the battery is dead before engine start, you're in trouble.

Hand-propping the aircraft, or using a GPU, may start the engine, but the alternator will never produce current without at least 2 amps from the battery, and you can't recharge the battery if the alternator isn't working. It is the classic Catch 22. Of course, the engine will continue to run because the ignition system is powered by the independent magneto system, but there will be no other electrical power in the aircraft.

Ammeters

The ammeter is one of the most helpful diagnostic tools. Pilots should become familiar with the type of ammeter in their aircraft and how it reacts to battery discharge, alternator failure, and loads that exceed alternator capacity. The alternator will provide very different information, depending on where it is placed in the electrical system. One thing is for sure: current flow means heat, and excessive current flow can generate enough heat to cause a fire. The pilot should be aware of trends in ammeter readings to head off a potential problem.

It is the role of the ammeter to provide information about current flow. The two types of ammeters are illustrated in Figure 8-13. They are the charge/discharge and the loadmeter. The charge/discharge or zero center type ammeter displays information about current flow. If the needle is to the right of zero, the alternator is working and supplying power to the electrical system. If the needle is to the left of zero, then the battery is discharging, indicating that the alternator is not supplying power to the electrical system.

The loadmeter or zero-left type of ammeter displays actual current draw (system demand) from the alternator. If the loadmeter reads zero, then the alternator is not supplying power to the system, leaving the battery as the sole source of power.

Whenever alternator failure occurs in flight, all operating electrical equipment will begin to deplete the battery. If that happens, the pilot must immediately assess the situation to determine what equipment is absolutely essential to the safety of flight at that moment and turn off everything else to lengthen the time to battery failure. This procedure, known as *load shedding*, will be explained later in the chapter.

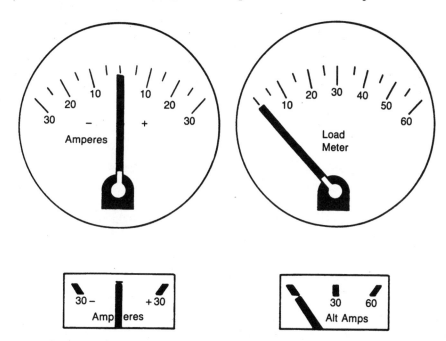

Fig. 8-13. *Types of ammeters.*

VOLTAGE REGULATORS

It is the job of the voltage regulator to make sure that the electrical system has a constant, regulated source of voltage, regardless of the condition of flight. Fundamentally, it is the electrical system's equivalent to a propeller governor. There is a rule of thumb that says if a battery is using too much water, the voltage regulator is set too high; if the battery doesn't seem to stay charged, then the voltage regulator is set too low. Essentially that is true. However, once again Murphy sneaks into the picture. In flight the battery is recharged by the alternator; this happens because the mechanic has set the voltage regulator to a value slightly above battery voltage. A 12-volt battery system will have an alternator output of 14.25 volts; a 24-volt battery system will have an alternator output between 28 and 28.5 volts. Because the alternator is putting out a higher voltage, it is in essence force-feeding the battery.

Unfortunately, the voltage regulator makes mistakes too. If it is set too high, the battery will overcharge, causing excessive operating temperatures, which means the active material will slough off all the faster, significantly shortening the life of the battery. Additionally, the hotter temperatures will cause the water in the electrolyte to evaporate, necessitating frequent additions of water. In general, having to add water more frequently than every 50 hours is an indication.

In extreme cases, the battery temperature may get so high that the plates will buckle or the electrolyte will boil and pour out of the battery vent caps. If, however, the voltage regulator is set too low, the battery will never completely charge. This is a tricky area in which to be playing detective, though, because if you have added electrical equipment to your aircraft, it is possible that the alternator is inadequate to meet the demand. The result would be the battery has to make up the difference, and instead of being recharged in flight, it would end up discharging by trying to share the load of the alternator. In this case one set of symptoms could indicate two very different problems: inadequate alternator and low-voltage regulator setting. Playing with voltage regulators can be deceptive, so consult a mechanic if you suspect a problem.

The Voltmeter

An excessive system-voltage reading on the voltmeter probably indicates regulator trouble. This can lead to several serious problems: The battery can be overcharged, causing severe, internal damage; acid may be forced out of the battery and damage surrounding equipment; and electrical overvoltage may damage other equipment, including light bulbs.

Overvoltage Protection

Some aircraft manufacturers incorporate an overvoltage control system. This protects the charging circuit from malfunction with a mechanical relay and solid-state triggering device. If the voltage reaches a preset value, the relay opens and the alternator field circuit disconnects. The relay remains open until the alternator switch is turned off. The electrical system then is supported entirely by the battery unless the aircraft happens to have a backup alternator installed.

Master Switch

There are several reasons why aircraft have a battery master switch. Some equipment, such as electrically driven gyros, do not have on-off switches. The pilot must be able to isolate them from the battery to prevent draining it when the engine is shut down.

Practicality dictates a convenient method of removing all electrical equipment from the battery simultaneously rather than turning each unit off individually. Safety dictates that during certain situations, such as emergency landings, the pilot should be able to shut down the electrical system to minimize potential fires. Logic says it really isn't a good idea to have wires running through the cockpit carrying potentially high amperage, so the actual master switch is powered by low-amperage current. It is this low-amperage current that operates an electromagnetic remote relay, located as close as possible to the actual battery, that physically connects and disconnects the battery from the electrical system.

If the alternator becomes inoperative, its exciter field will continue to demand system current, which is now supplied by the battery. To solve this problem, manufacturers have devised a "split" master. One-half of the master switch controls the alternator and the other half the battery. It is possible to remove the alternator from the electrical system and still use the battery. The rule is, if the alternator fails, turn off that half of the master switch.

Alternator Preflight

If practical, a preflight inspection should include a good look at the alternator. Mounting bolts should be tight and clean because they form an electrical connection to the aircraft. In fact, they should be tightened to specific torque value: too loose and the unit vibrates and shifts in its mounting, too tight and the lugs and/or brackets may crack or break.

The drive belt, which also must be properly torqued, should not be so loose as to allow slippage and loss of alternator output. On the other hand, if it is too tight, the belt may break; even worse, it may cause a side load on the alternator shaft, imposing an abnormal load on the bearings and seals. Such a condition would lead to an early failure of the unit. If you can see the alternator fan, visually inspect it for general condition. It should not have any cracks, something to watch for particularly in the area of the welds. The fan should not appear bent, and it should have sufficient clearance to turn without scraping a baffle or other structure.

In multiengine aircraft one of the most common complaints about the dual alternator system is an imbalanced output reading. If both alternators use a single regulator, it is neither uncommon nor inappropriate for an imbalanced output reading to occur. The old dual generator systems needed a balanced output to prevent component damage, so over the years many pilots became accustomed to watching for an imbalanced output reading. Old habits die hard. Output readings on modern dual alternator systems do not have to be balanced. Variations in resistance characteristics between two charging circuits will cause imbalanced indications. This could be the result of voltage drops in the charging system, wiring, or ground-circuit connections or even variation in manufacturers' tolerances between alternators.

For instance, if the load requirement from two systems were 60 amps, the voltage regulator would allow sufficient current through the field circuits to meet that requirement. The charging circuit with the lesser amount of resistance will use more current than the other, but between the two, they will produce the necessary 60 amps, and that's all that matters.

If the output fluctuates or there is no output at all, the most likely cause is a loose or broken alternator belt. If you recently have added an alternator or made some related system change and you find that the aircraft lights flicker, you detect poor alternator voltage regulation, you see the instruments oscillate, or you hear radio interference, you may have an excessive voltage drop. The most probable causes are improper conductor diameter or length. Aircraft with alternators are more prone to have radio interference because of the constant change of voltage and current signals conducted along and radiating from conventional wiring. These fluctuations are picked up by the avionics equipment either through their power source or antenna wires. The static almost always can be reduced dramatically, if not eliminated completely, by the strategic placement of a capacitor and installation of shielded conductors between the regulator and the alternator.

ELECTRICAL GROUNDING

Those of you who like to poke around inside the aircraft, more than that afforded by just a preflight, probably have already noticed a significant difference between the way your household electricity is distributed and aircraft power distribution systems. If you look closely at the cord leading to your table lamp, you will notice there are actually two wires, bound together, running from the lamp to the power source (wall outlet). Then take a look at an aircraft component; there is only one wire.

Years ago when aircraft covered with fabric were standard, the electrical systems had two wires. Today's modern, all-metal fuselage allows the airframe to act as one of the conductors, eliminating one of the wires. A single wire supplies electricity to the component, which is grounded to the airframe. The battery is also grounded to the airframe, completing the circuit. The advantages of eliminating half the wiring are obvious: lighter weight, less complexity, lower construction costs, and reduced potential problems.

THE ELECTRICAL BUS

All power distribution points, also called buses, are essentially the same. Figure 8-14 shows that a bus is a metal strip to which parallel circuits are connected. They are typically named for their source of power or the function they serve. For instance, a bus that is directly connected to the battery through the master switch is usually called the battery bus. One that is always connected to the battery with no method of disconnection is a hot battery, or essential, bus because the bus is always "hot" with electricity.

Items on a hot battery bus will deplete the battery even when the master switch is turned off. Therefore, it is important to assure that all such equipment is individually turned off. While most commonly found on larger aircraft, a limited hot battery bus is installed in some lighter aircraft. The Beech 58 Baron uses one to activate baggage and courtesy lights and the stall indicator. The Cessna 172RG uses it to run a clock and flight hour recorder.

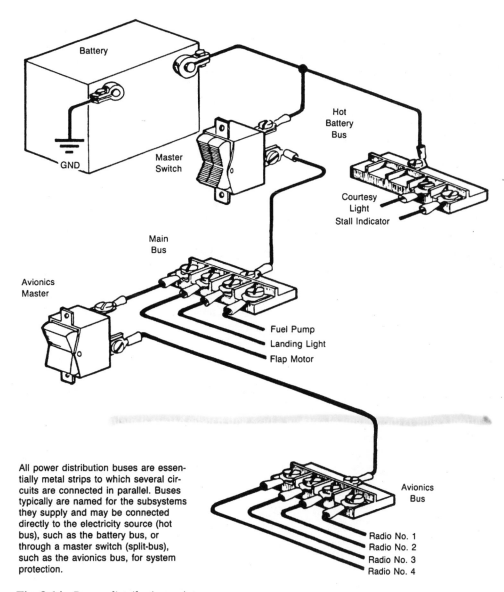

Fig. 8-14. *Power distribution points.*

The Avionics Bus

The avionics bus, which is designed as a split bus system to allow separation of the avionics from the rest of the electrical system, is another example of a bus named for its function. The manufacturer typically uses one of two means to remove it from the battery bus. One method is an avionics master switch that must be turned on and off by the pilot to activate the avionics package. The other method is an automatic relay, which is usually designed so that whenever external power is applied, or the starter is activated, the avionics bus is isolated from the airplane's electrical system automatically. This prevents potential power spikes from harming the avionics during engine start or when an external power source is hooked up to the airplane.

CIRCUIT PROTECTION

When scientists first began experimenting with electricity, there was a need for a weak link in the system to protect both the scientist and the test equipment from being electrocuted. Initially, somewhere along the circuit there was an open wire, generally undersized and uninsulated. Since an overload causes an increase in the wire temperature, the undersized wire would melt before the rest of the system. This weak, or fusible, link was eventually shortened to the name fuse.

The Fuse

The primary purpose of a fuse in an aircraft is to protect the system wiring. The secondary purpose is to protect a given appliance. In the modern fuse, higher-than-normal current heats up the fusible element until it reaches its melting point, then the fuse blows. The blown fuse is simply pulled out and replaced with a new one, and it is always a good idea to keep an extra set of replaceable fuses in the aircraft. In choosing a replacement fuse, it is important to consider three specifications: current rating, voltage rating, and fusing characteristics.

Current Rating

Current rating is the amount of current a fuse will handle before it melts. What is perhaps not so obvious is the effect that ambient temperature has on that rating. Since any given fuse will melt at some specific, predetermined temperature, it stands to reason that however that temperature is achieved, the fuse will melt. While it is unlikely that the ambient temperature will ever reach the fuse-melting temperature without the presence of a fire, any temperature increase will increase the rate of failure. The reason for this is that during periods of high ambient temperature (such as on hot days), the fuse starts out at a higher temperature before the current is ever applied to it. Therefore, at an ambient temperature of approximately 80 degrees F, the fuse ampere rating should be 25% higher than the normal operating current of the circuit.

Voltage Rating

Perhaps most confusing of all is the voltage rating. If you look at a fuse you will find one of the following voltages stamped on it: 32, 125, or 250 volts. These ratings indicate the maximum voltage for which the fuse is usable. If the fuse has no rating, you may assume it is 32 volts. The 32-volt fuse is usable for all aircraft DC systems (14 or 28 volt). The 125-volt rating could be used for 14 or 28 VDC, or 115 VAC, and so on. Therefore, fuses should be rated equal to, or greater than, the voltage of the circuit or equipment.

Fusing Characteristics

The third rating, fusing characteristics, deals with the speed at which a fuse will break the circuit. There are two basic categories: normal blow and slow blow. The normal blow fuse may be further divided into fast acting and medium acting. These types of fuses would be used in circuits where no surges or transient voltages are expected. If a circuit is normally subjected to transient voltage or surges such as engine starting loads, then the slow blow fuse should be utilized. Incorporating a built-in time delay, the slow blow fuse will not immediately fail if the maximum temperature is reached for a short period of time.

Circuit Breakers

The circuit breaker (CB) is a newer type of circuit protection. In principle, the CB performs the same task as the replaceable fuse but has the convenience of being able to be reset. In the CB the element also reacts to heat, but rather than melting, it expands, causing the circuit to open (trip). Approximately two minutes after it trips it will have sufficiently cooled down so that the pilot may reset it. Resetting a CB is a simple matter of pushing it back to its original position.

When a CB rated under 20 amps trips, it is typically considered a "nuisance" trip and is simply reset. Nuisance trip means the reason for it tripping could be a host of transient reasons, none of which are a real problem. CBs with a rating over 20 amps are normally not considered a nuisance trip and should not be reset. It is assumed there is a problem in that case.

The only real problem with this type of CB is that because they are flush mounted, it is impossible to intentionally deactivate an entire circuit. All equipment must be shut off individually. One note of caution: Never hold a CB in if it wants to pop; that will almost certainly lead to further damage of the circuit and a probable fire.

CHANGING FROM DC TO AC

While the aircraft electrical system utilizes DC voltage, some equipment requires 26 or 115 VAC. To accommodate this, an inverter is used to change DC into AC. In light, single, and multiengine aircraft, the need for AC is very limited, so an actual inverter is seldom used. Rather, any equipment that requires AC voltage has a type of miniature inverter in the instrument, greatly reducing the cost and weight associated with installing an actual inverter.

STARTERS

A smoothly running engine is about as close to perpetual motion as you can get. It automatically meters the correct amount of fuel; times the ignition, intake, and exhaust valves; and produces sufficient electrical energy to continue running. One thing it cannot do is start itself. Some force must rotate the engine to make the valves, fuel flow, ignition, and magnetos begin operation. The starter—sometimes referred to as a cranking motor—is simply a battery-powered motor that rotates the crankshaft fast enough to get the engine started. The basic operating principle is magnetism.

An electric current-carrying conductor, such as a wire, tends to move when placed in a magnetic field. In Figure 8-15, the current is moving away from the reader. This is indicated by an X, which represents the tail of an arrow pointing in the direction of travel. Whenever there is moving electric current, there also is a magnetic field surrounding it. Using accepted electrical theory, it can be determined that in this case the magnetic field around the conductor is moving counterclockwise.

The direction of the magnetic lines around the main field—the north and south poles of the magnet surrounding the wire—move from north to south (top to bottom). The magnetic lines of the two fields on the right side are in opposition and effectively cancel each other. On the left side, the two fields travel in the same direction; the effect

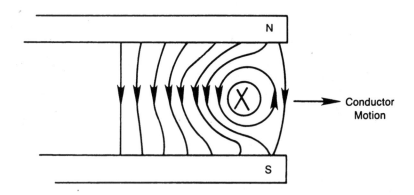

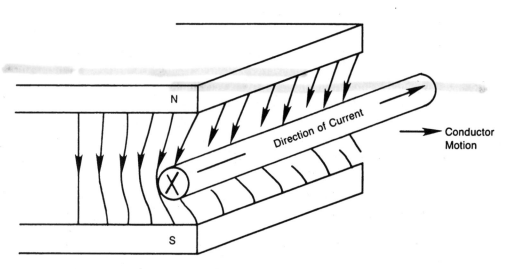

Fig. 8-15. *Interaction of magnetic forces.*

is to create a stronger magnetic field. With a strong field on its left and a weak on its right, the conductor moves toward the right.

A single conductor passing through a magnet isn't a very effective tool. If we bend the conductor into a coil so it comes back through the magnet, we have a rudimentary motor. With the current applied in the direction of the arrows, the forces of the magnetic fields tend to push up the left side of the coil and push down the right side.

By attaching the coil to a freewheeling shaft, it will rotate in a clockwise direction as shown in Figure 8-16. This primitive, low-power motor can be somewhat enhanced by using heavy copper wire for the coil, which permits heavier current flow and a stronger magnetic field. Similarly, if current is run through windings around the magnetic poles, called *field windings*, there also will be a stronger magnetic field between them. Our single-coil system still would operate poorly because the torque and speed would vary dramatically as the coil moved from high magnetic influence to low.

The solution is a motor armature composed of many insulated coils connected to an iron or steel core. The core, mounted on a shaft with bearings at each end, increases and focuses the strength of the magnetic field to maximize its use. As one coil moves out of the magnetic field, another coil enters it. The armature turns continuously and smoothly as long as current is applied.

The starting circuit system includes the starter motor, switch, battery, and load circuit. The battery powers the starter motor and operates electrical equipment whenever the aircraft generator is unable; therefore, a healthy battery is essential to effective starter operation. The starter switch activates and deactivates the starter motor. It completes the circuit between battery and motor. Load circuit is the term used for all the cables that connect the individual units of the starter system, including the battery ground strap, the battery-to-starter switch, and the starter switch to motor.

The load circuit must be of sufficient capacity to carry the necessary starting current with minimal loss to resistance. These cables are chosen carefully to handle high current; a smaller size causes significant reduction in cranking power and can be a fire hazard. Cable connection points should be inspected periodically to assure they are tight and clean because any unnecessary resistance also will reduce cranking power.

The starter motor converts battery electrical potential to mechanical rotary power. Its frame and field assembly house and support all motor components. The field windings and pole shoes combine with the metal frame assembly to provide a path for the magnetic field in which the armature will turn, as shown in Figure 8-17. A simple but effective brush and holder system feeds the rotating armature with battery power. The brushes seat against, and slide across, the commutator as it turns. Each segment of the copper commutator is insulated from the others and the armature shaft, permitting battery power to go to each individual coil. The armature bearings are seated on the end with the commutator end head, and on the other end in a pinion housing, which also contains the gear drive mechanism that turns the engine.

Various airframe manufacturers employ different methods of activating the starting circuit. Many old aircraft have a T-handle to pull or a solenoid button to push. Most modern aircraft have a key switch similar to an automobile. When the switch is turned, battery current goes to the starting motor terminal, which divides it between the field

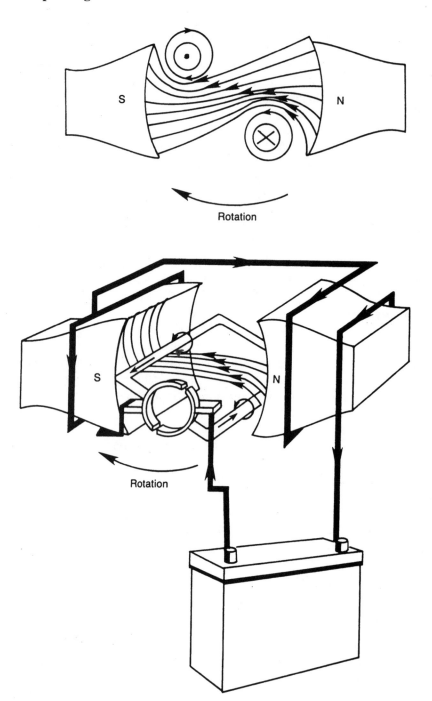

Fig. 8-16. *Interaction of magnetic forces on a coil.*

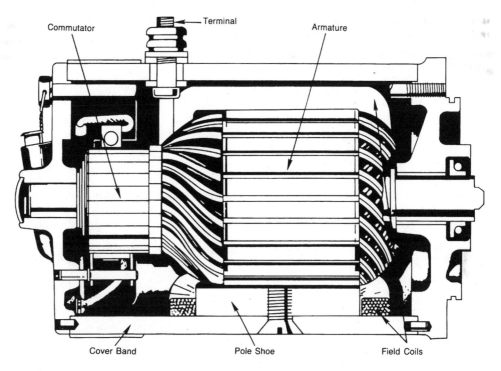

Fig. 8-17. *Aircraft starter motor.*

windings around the magnet and the brushes to the armature windings. In both cases, it then goes to ground. The simultaneous current flow through armature and field windings results in a strong magnetic field, which makes the armature turn.

The armature is connected to a drive system, which engages and disengages the motor to the engine flywheel. If you think about how a bicycle works, you will understand why a gear reduction is placed between the starter motor and the flywheel. Without gear reduction, the little motor would have difficulty turning the large, high-compression engine fast enough to get it started.

Starter Drive Mechanisms

The two principle types of drive units are the Bendix drive and the overrunning clutch. The Bendix unit operates by a combination of screw action and inertia. While there is some variation between systems, essentially what happens is the starting motor is engaged, which causes the armature and its pinion gear to rotate. The screw action forces the pinion gear forward and it meshes with the engine's flywheel ring gear. After the engine starts, the flywheel begins to turn faster than the armature, so the pinion gear accelerates and is threaded back along a specially designed shaft, out of mesh with the flywheel.

The pinion gear of the overrunning clutch usually is shifted manually, such as with the old Cessna 150 T-handle; however, on some models it is controlled by a solenoid. As the gear moves forward, it engages the flywheel and the engine starts. The clutch then re-leases and the pinion retracts. To assure the pinion does not harm the starter by turning too fast when the engine starts, it is designed to be able to turn faster than the armature without damaging it.

In either type of drive system, if the pinion gear cannot disengage for any reason, the engine may drive the starter motor, causing it to burn out. The pilot should never knowingly allow this to happen, and engine shutdown is the only solution. Some manufacturers put an annunciator light on the instrument panel to alert the pilot of starter disconnect failure.

LOAD SHEDDING

It is a cloudy, wintry night as you fly home after a much-needed weekend getaway with the family. As the airplane momentarily slips out of the clouds, you catch a glimpse of the stars, but the ground remains a mystery beneath the soft, billowy, low overcast. Your home airport, still an hour away, is reporting weather that should mean an uneventful instrument approach. The hum of the engine combined with the silent efficiency of your autopilot gives assurance that you are at peace with the world.

As you scan the glowing instruments, you can see the reflections of your napping family. Suddenly, the sleeping faces are bathed in red light. The alternator-out light has illuminated; from now on, the only source of electrical power is the battery. You need to reduce electrical load to the bare minimum (called load-shedding). What you do in the next few minutes will make the difference between a flight you will long reminisce about during hangar flying sessions and one that could terminate in disaster.

Someone once said, "Man's flight through life is sustained by the power of his knowledge," and nowhere is that more true than in the case of an alternator failure dur-ing single-engine, night-IFR operations. However, before multiengine pilots stop reading this, let me pass on a little story.

In a recent discussion with the pilot of a medium-sized corporate twin, I was extolling the virtues of having two engines, and therefore, two alternators. I pointed out the incred-ible odds against ever suffering total electrical-system failure. He smiled and mentioned that the previous month both alternator clutches failed simultaneously; fortunately, it hap-pened while he was still on the ground. Two-engine types, beware; it is not outside the realm of possibility.

During the days when aircraft had few or no electrical systems, more often than not, the pilot was also the mechanic. Airplanes were mechanically simple and the solution to a given problem was fairly obvious. As demand increased for more sophisticated sys-tems, being a pilot began to require increased development of flying skills, leaving less time to devote to mechanical familiarity.

It is not hard to imagine that as electrical systems became more involved, mechanics began sketching them out before actually wiring the airplane. The more complex the sys-tems became, the more careful the planning and the more elaborate the schematic draw-ings. Because these schematics were intended to be used by engineers and mechanics, no attempt was made at realistic depiction of components. Ultimately systems became so complex, problems could arise that were no longer easily solved—or even understood.

Pilots became aware of the need for increased systems training to be able to troubleshoot problems in flight. Because schematics had become the key to unraveling the mysteries of any given system—not just electrical—pilots had to learn to interpret them. Unfortunately, systems continued to become more complex and schematics rapidly took on nightmarish qualities. I cringe when I think of the electricity course I had to take as a maintenance student at the University of Illinois, and the hours spent poring over incomprehensible electrical-system schematics.

Electrical-System Schematics

The purpose of a schematic is to provide a means for the mechanic to trace a system visually. It is typically used as an aid in locating the source of a problem, and essentially is a road map for troubleshooting. Unfortunately, maintenance schematics go into far greater detail than is required by the pilot for whom only that information, which may be used to solve in-flight problems, is useful.

Thanks to the General Aviation Manufacturer's Association (GAMA), simplified pictorial schematics were initiated some years ago for general-aviation aircraft. The effect was a quantum leap forward in pilot understanding and ability to troubleshoot systems. Gone were the obscure symbols for motors, alternators, starters, switches, and other components. Everything was replaced by simplified, miniature drawings of actual components, as shown in Figure 8-18. Alternators now look like little alternators, master switches like little master switches. Fuses, circuit breakers (CB), switches, and other controls and equipment are easily distinguished and diagrammed in a simplified, functional order that enhances system understanding. The modern pilot-oriented schematic permits good system understanding by even the most mechanically unsophisticated pilot.

When I was a King Air simulator instructor for FlightSafety International, I was introduced to a game called "what if," which I subsequently taught to all my crews. Designed to keep a pilot sharp on systems and emergency procedures, "what if" can be played anywhere. If there are two or more players, one person asks a question and the others see if they can answer it. For instance, what if the gear won't extend even with the manual gear-extension procedure? Or what if the alternator field CB pops? If you are playing alone, then sit in your airplane with the manual close at hand and ask yourself "what if" questions, referring as necessary to the procedures and system section of the manual.

To be able to play "what if" successfully, a pilot must be familiar with the systems and procedures of the airplane. "What if-ing" the electrical system requires not only a basic understanding of the system, but also what specific equipment it operates. Radios, lights, and pitot heat are obvious, but there are some not-so-obvious ones. Some equipment may or may not use electricity: for example, hydraulic landing gear, fuel valves, gyros, and stall-warning devices.

On the other hand, some instruments that you might expect to be powered by the electrical system may not be, such as cylinder head temperature, oil pressure, and oil temperature gauges. The point of the game is to keep an in-depth knowledge of your aircraft systems fresh in your mind.

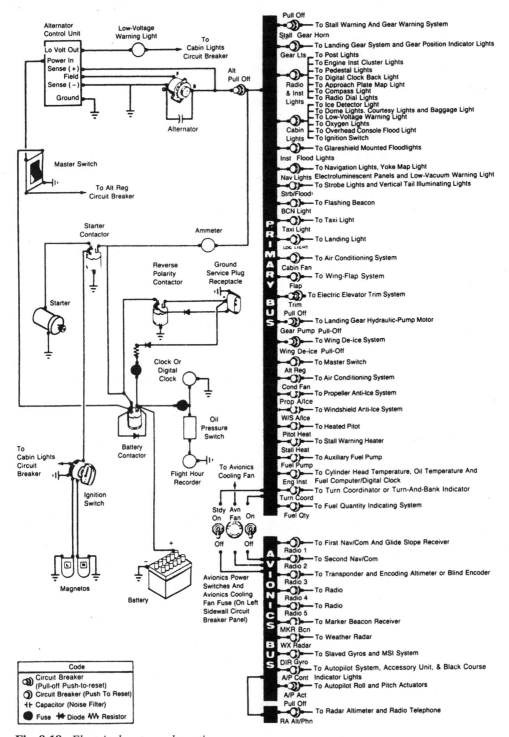

Fig. 8-18. *Electrical-system schematic.*

System Failure

Electrical-system failure is a very commonly misused phrase that almost always actually refers to alternator or voltage-regulator failure. In such cases, what happens is the alternator is no longer capable of providing electricity to the electrical system and the battery takes charge. The electrical system still functions, but at the cost of an ever-discharging battery. There is no way to recharge the battery, and if the situation continues long enough a real electrical-system failure may occur. Actual instances of total electrical system-failures are almost nonexistent.

By way of introduction to the subject of alternator failure, I used to ask my university ground school classes to tell me what they felt the average pilot thinks about alternator failure. There would usually be at least one student who would quickly respond, "They know it's not a problem on day-VFR flights." The observation was good; unfortunately, the premise was not.

Contrary to popular belief, all in-flight alternator failures are bad. I will agree there are degrees of bad, and probably the worst condition would be night IFR flight in icing conditions, but day VFR also can present some sticky situations. Besides avionics equipment, there are landing gear and flaps to consider. True, you can usually manually extend the gear, but what happens if you have to go around and there is no power to retract? Night flight presents another major problem—the simple challenge of being able to see whatever instruments still happen to be working, for instance.

I had an instrument student who was of the opinion that if the lights went out in flight, he would simply hold his flashlight in his mouth. While on a dual flight one night, I turned off the battery master and informed him that he had "lost" the electrical system. Smugly, he reached over and pulled a flashlight out of his flight bag. It was one of those nice, big, bright, ribbed chrome types. It just barely fit in his mouth, took about one minute to make his jaws ache, three minutes to make his head stoop, almost knocked out a tooth when we hit a little bit of turbulence, and the light intensity was blinding as it reflected off the instruments and windshield. Could you imagine doing that for an hour or two? So much for holding a flashlight in your mouth.

The simple truth, which all pilots must be willing to face, is you have to be prepared for the possibility of alternator failure. What makes this failure such a problem is the electrical system continues to operate normally as if nothing had happened. When the alternator fails, the battery immediately takes over the power requirements of the system. Due to the failure of the alternator, the battery cannot get recharged, so any electrical demand will cause the battery to discharge. The greater the demand, the faster the battery will become exhausted. The trick is to identify the problem immediately, while the battery is still fully charged, so you can assure sufficient power for the remainder of the flight. That's sometimes easier said than done. If the electrical system itself doesn't show any change initially, how does a pilot know when an alternator failure occurs?

Alternator Troubleshooting

The prime source of information regarding alternator failure is the ammeter; the telltale indication depends on which of the two types you have in the airplane. As I said earlier, the zero-center type of ammeter shows charge/discharge rate; when it shows a discharge you may assume the alternator is inoperative. The zero-left, or loadmeter type, indicates the amount of load the alternator is supplying. A zero indication of the loadmeter indicates the alternator is inoperative.

The real problem is that the ammeter is seldom in the pilot's normal instrument scan and often is installed on the far side of the instrument panel or somewhere below the pilot's normal field of vision and obscured by the yoke. Many aircraft have a red warning light to indicate alternator failure, which is certainly helpful, but bright daylight conditions and preoccupation with other problems both have been known to distract a pilot sufficiently enough for the light to be overlooked. The safest course of action is to learn to include the ammeter in your instrument scan.

It is important to point out that an inoperative alternator does not necessarily indicate alternator failure. In the event that you have an inoperative alternator indication, before taking any drastic measures such as load shedding or early termination of the flight, do a little snooping around. It is possible that the master switch may have been shut off accidentally or the alternator (or alternator field) CB may have tripped and a reset attempt may be successful. In aircraft that have the on/off-type CB installed for the alternator, it may have been accidentally shut off.

A good rule of thumb in all troubleshooting is to initially ask yourself what is the most obvious possibility for the failure? Often the answer will be miscontrol of the system, such as accidental deactivation. While it is particularly distasteful to accept, it is nonetheless very commonplace for pilots to turn off master switches instead of turning off lights, switch to empty fuel tanks instead of full ones, and raise landing gear after touchdown instead of flaps. Human error is something all pilots must watch out for.

Most pilots will agree at this point that in-flight alternator failure is a nasty situation at best. While the importance of early recognition hardly can be disputed, it does not assure the safe outcome of the flight. It is early recognition combined with knowledge of alternatives that will assure the best possible decisions under the circumstances.

Load-Shedding Decisions

To make appropriate load-shedding decisions, the pilot should study the electrical system schematic while on the ground to get an understanding of how the system is designed and what can go wrong. The more complex the system, the greater the need for digging in and finding out what makes it tick, but even a relatively simple system merits a thorough review and some research.

Looking again at the sample schematic, it is possible to quickly note several important details. This system has a "split master," which allows the pilot to easily remove the alternator field from the system. This is important because even if the alternator fails, the alternator field will continue to demand precious current from the battery. Therefore when the alternator fails, the pilot should shut off that portion of the master switch. It also shows which items have CBs that permit the pilot to remove the item from the primary bus. They include the landing gear hydraulic-pump motor, strobe lights, and alternator as depicted in the illustration.

In addition, there is an avionics power switch, (avionics master) that permits separation of all avionics from the primary bus, a valuable practice during engine start to reduce the risk of damaging avionics from possible, potentially harmful voltage fluctuations. Another useful piece of information is exactly what items are protected by each CB. If, for instance, the CB marked NAV LT trips, the problem could be rooted in a few possible areas: the navigation lights, the control-wheel map light, the electroluminescent panels, the low-vacuum warning light, the wiring to any of those items, or the CB itself. Experimentation will produce sufficient information quickly enough for you to make intelligent decisions.

First, turn off all the items under the control of that circuit breaker, then, after a two-minute cooling period, reset the CB by pushing it in. Wait for a minute or two and if it pops again, the problem is either in the wiring or the CB itself, both of which are beyond the pilot's in-flight capability to repair. If nothing happens, activate one of the systems and wait another minute or two. If still nothing happens, turn it off, activate a different system, and wait. Chances are pretty good that eventually you will activate the faulty system, which will cause the CB to trip a second time.

It is worth noting that a glitch in a system occasionally will cause a CB to trip, and upon troubleshooting the CB will not trip a second time—so much the better. This procedure should be followed whenever there's a trip of any CB that protects more than one piece of equipment. One item of information that is essential to making intelligent load shedding decisions, which is omitted from many schematics, is what constitutes essential equipment.

"Nonessential equipment," as it applies to the electrical system, refers to equipment that should be turned off in the event of alternator failure. CFIs, when discussing alternator failure, tend to say, "...and of course then you turn off all nonessential equipment." Unfortunately, many CFIs never actually get around to telling you how to determine what is nonessential.

Certainly, if you are flying along listening to stereo music and have a portable coffee maker plugged into the cigarette lighter outlet, there is some fairly obvious nonessential equipment being used. How about the situation where you are flying IFR at night and the closest airport with weather good enough for an approach is more than an hour away? The solution to the problem is to manipulate usage of the electrical equipment to assure you will not exceed the capability of the battery for the time you anticipate using it. That means you will have to begin thinking of each piece of equipment in terms of its necessity at any given moment, as well as how much amperage it will require to use it.

Take, for instance, a battery rated at 40 amps per hour. Theoretically, it will support a continuous system load of 40 amps for one hour. If you expect your flight to last one hour, you have 40 amps at your disposal during that hour. That implies you may use whatever equipment you want, provided you never exceed a total of 40 amps demand at any given time for one hour.

I wonder if Charles Darwin was thinking of batteries when he said, "There is no subject, however complex, that with patient and intelligent thought, will not become even more complex." Remember that this is purely theoretical; in reality things just don't work that neatly. For one thing, it presupposes the battery was fully charged and in perfect condition. It also does not take into consideration that some equipment, such as radios, may not work as the battery becomes significantly discharged even before the anticipated time expires. Therefore, in this case, it is not practical to think that reducing system demand to 40 amps will guarantee one hour of use. It is always a good idea to reduce the load as much as is safely possible and to terminate the flight as soon as you can.

Developing a Load-Shedding Chart

To make intelligent decisions about what equipment to use and what to turn off, the pilot needs to do a bit of homework. One afternoon when you have the bug to fly but the weather won't be agreeable, take the electrical-system schematic to the airplane, find the current draw for each item listed, and develop an electrical-emergency load-shedding list like the one depicted in Table 8-6 for a common general-aviation single-engine aircraft. This can be done in a number of ways.

Perhaps the easiest way to develop your list is to check the CB rating for each item. While it will be higher than the actual current draw for that particular item, approximately 1.1 to 1.5 times higher, it will give you a pretty good estimate on the safe side. A more exact method, though more time-consuming, is to get the rated amperage from the data plate on each appliance. The same information can also be found on manufacturer's information sheets. Compare the differences in the CB amperage ratings shown in the

Table 8-6. Electrical Emergency Load-Shedding List

Amps	Item
20	Taxi and landing lights
20	Landing gear hydraulic-pump motor
10	Pitot heater
10	Wing-flap system
10	Flashing beacon
5	Auxiliary fuel pump
5	Avionics cooling fan and strobe lights
5	Instrument cluster Low voltage warning light Ignition switch
5	Carburetor air temperature gauge, map, compass, instrument lights and dimmers, dome and courtesy lights, post lights
5	Navigational lights Yoke map light
5	Turn coordinator
5	Radio 1
5	Radio 2

electrical emergency load-shedding list and the list of actual rated amperage for the same airplane in Table 8-7. It is worth noting that some older communications radios may require up to two amps to receive and six amps to transmit.

Once you have studied the system and have your electrical emergency load-shedding list, then you are prepared to make intelligent decisions should you be forced to rely only on battery power. It is easy to see, for instance, that taxi and landing lights should be avoided and the gear should not be extended until you are absolutely sure you will be landing, as the price for gear extension and then retraction is very high indeed. Take the situation where you are about to fly an instrument approach to an airport with marginal weather. If there is the likelihood of a missed approach, you might seriously consider manually extending the gear to conserve battery power for a possible trip to the alternate. In most aircraft, the gear then can be electrically retracted in the event of a missed approach.

One important lesson to be learned by all of this digging and probing is that the average battery in good condition should provide sufficient power to run necessary electrical equipment long enough to get to a suitable airport. Not only is that fact important from an operational standpoint but perhaps even more important from a psychological one.

PREVENTIVE MAINTENANCE

In addition to the few approved preventive maintenance items that pertain to the electrical system in Appendix A to Part 43—Major Alterations, Major Repairs, and Preventive Maintenance—there are other things you can do that will help prevent problems. Keep the battery clean, charged, and correctly filled. Always check ground power units for proper voltage and polarity before hooking up to the airplane. Check the security, cleanliness, and condition of wiring that can be seen during preflight. Discipline yourself to include the ammeter in your normal instrument scan, and perhaps the best advice is to generally be aware of the state of the electrical system.

Lead Acid Battery

The key to long and happy lead acid battery life is also good preventive maintenance. Following these nine principles will go a long way toward keeping your battery in top condition.

1. Keep the battery fully charged at all times.
2. Never hit battery posts with a hammer in an attempt to remove terminal connections.
3. Keep the electrolyte level up by adding water as necessary.
4. When not in use for extended periods, store the battery in a cool, dry area and recharge it at least every five weeks.
5. Check terminal connections for corrosion, which CAN be removed by gently brushing the surface with a hard bristle brush, and apply Vaseline to the studs and terminals to prevent further corrosion.
6. Never pry cable connectors with a screwdriver in an attempt to remove them.

Table 8-7. List of Actual Rated Amperage

Amps	Item
17.5	Landing gear motor
9.0	Taxi light
9.0	Landing light
8.5	Wing-flap system
6.0	Flashing beacon
3.0	Strobe light
2.5	Autopilot
2.25	Com (transmitting)
2.0	Transponder
1.2	DME
1.2	Dome and courtesy lights
1.0	Avionics cooling fan
1.0	Navacom (receiving)
1.0	ADF
0.7	Map, compass, and instrument lights
0.5	Glideslope
0.1	Marker beacons

7. Never overtighten terminal bolts; you unnecessarily jar the battery post.

8. Never reverse battery leads; positive goes to positive, negative to negative.

9. Assure that the battery is securely held down without excessive tightness. If the straps are too tight, you risk damaging the battery container.

NiCd Battery

Most manufacturers consider it good preventive maintenance to check new batteries every 50 hours for the first few months of operation. After a while, a pattern will emerge and the time between further checks can be adjusted accordingly. Whenever working with an NiCd battery, it is extremely important to remove jewelry and any metal articles from your body. Touching points of opposite polarity can literally weld the article to the battery and cause injury.

Electrolyte is extremely caustic and should never be left on aircraft skin or in the battery case. Always rinse and clean the area thoroughly after spillage. Obviously, it will also burn clothing and skin very quickly. Always wear protective clothing when servicing an NiCd battery. This should include at least rubber gloves, rubber apron, and protective goggles. If electrolyte should ever get into your eyes, quickly flush them with a large amount of water and get medical attention immediately. Skin burns should be treated by flushing the area quickly with water, then neutralizing the chemical with water and a 3% solution of acetic acid, vinegar, or lemon juice. A 10% solution of boric acid also may be used. Be very cautious about battery fumes. Work only in a well-ventilated area. An overcharged battery gives off hydrogen and oxygen gases, which together can be explosive. For that reason, make sure battery terminals are tight to prevent sparking.

The hydrometer, a useful tool in determining the state of charge in the lead-acid battery, is useless in testing the NiCd battery. NiCd electrolyte shows no density or composition change during the charge/discharge process, so it is impossible to use electrolyte-specific gravity, fluid level, or even battery voltage to determine the charge of the battery. The only reliable way to determine a state of charge is to discharge the battery at a constant rate, according to the manufacturer's instructions, accurately timing how long it takes. An experienced mechanic can determine state of charge based on that information.

Whenever the battery is removed from the aircraft, it should be discharged according to the procedure outlined in the manufacturer's manual and then serviced. Carefully check for general condition. A light powder deposit of potassium carbonate is not uncommon on the top of the cells; simply remove and clean the deposits. Do not use petroleum spirits, trichloroethylene, or any other type of solvent, as they are harmful to the battery. Brushes also are not a good idea because they tend to force particles between the cells. Use a clean cloth to wipe both the battery case and cover. If there is particulate matter inside the case, use filtered, compressed air to blow it out and wear safety goggles! During the visual inspection, check for cell-vent integrity. There should be no obstructions or damage to the vent.

When inspecting the battery, if the cell electrolyte level is low, you may add either distilled or demineralized water. Never use tap water because the minerals will cause shorting of the plates. Never add water unless the battery is fully charged because the electrolyte level lowers as it is absorbed during discharge. When the battery is brought up to full charge, the level will raise. Approximately three hours after a battery has been charged, the electrolyte level should be 1/8th inch above the visible insert in the cell. If not, fill to that level.

If a battery is partially discharged when removed from the aircraft, you should charge it. If it has been inactive for more than two weeks but less than two months, it should be charged before attempting to use it again. An NiCd battery that has been inactive for more than two months should be reconditioned. This consists of completely discharging and recharging the battery according to the manufacturer's recommended procedure. Often referred to as "deep cycle," it is a preventive maintenance procedure that assures all the cells of the battery are pulling together. In addition, most problems relating to the battery's state of charge are solved by this reconditioning process.

It is difficult to fix a specific number of flight hours for preventive reconditioning. There are many factors that affect the frequency. According to the Operating and Service Manual published by General Electric for its NiCd battery, all of the following are considerations: how well the battery and aircraft electrical system is matched, battery and electrical component maintenance, geographic location and season of the year, operator techniques for engine starting, frequency, and severity of engine starts, and battery operating temperatures.

Sometimes a battery does not seem to supply its rated power. The probable cause is frequently discharging the battery at a rapid rate then quickly recharging it, which results in a cell voltage imbalance. The corrective action is to recondition the battery. In the absence of other information, have the battery checked every 50 flight hours until you have sufficient historical data to increase or decrease frequency. The range can be from a few hours to more than 1000!

Preflighting the NiCd battery is relatively simple. Check the box for structural and chemical damage. Vent and cooling lines should be checked for leaks, obstructions, or any other damage. If there is excessive electrolyte spewage or use of water by one or more cells, it may be the result of any one of several problems. The cell could have been overfilled or filled when the battery was in a state of discharge. The vent cap could be loose or there could be damage to the O-ring or vent cap. The voltage regulator charge voltage may be set too high, overcharging the battery. And finally, if a visual check doesn't reveal any problems, check the voltage regulator.

Alternator

The most significant preventive maintenance for your alternator is to assure proper polarity at all times. If you change batteries, use an external battery boost, or during a fast charge, always assure proper polarity. A mistake here can cause permanent damage to the alternator, its rectifier, and many electrical-system components. For the same reason, be very careful not to reverse regulator leads. It also is important to guard against transient voltages, which will damage semiconductors in the alternator, radios, and other electronics. A 140-volt transient voltage for more than .001 second will fry a voltage regulator! To help preclude this possibility, always keep the battery in the circuit; it serves as a sort of electrical shock absorber.

Old-style generators required "polarizing" when they were new to assure proper polarity. This was done by attaching a temporary external connection to the field circuit. Alternators NEVER require this; it will destroy the semiconductors. Obvious preventive maintenance items include a snug alternator belt and battery in good condition.

Perhaps not so obvious, the voltmeter and ammeter should be checked periodically to assure proper operation. The ammeter alone is one of the most important diagnostic tools of the alternator. Another frequently overlooked but very important preventive maintenance item is the general condition of the cowling baffles and airflow path. Alternator temperature limits are critical and are determined by the type of winding insulation, bearing lubricant, and type of semiconductors used. Even the voltage regulator and overvoltage control have temperature limitations. Baffles naturally take a beating as mechanic after mechanic remove the cowling and replace it. As baffles bend, deform, and crack, the airflow becomes altered and ultimately will lead to premature failure of the alternator as well as other parts.

Starter

The best preventive maintenance for the starter system is to use it as little as possible. Another life extender is to keep the engine in tip-top shape. Problems with the fuel, ignition, and lubrication systems result in a harder-starting engine, which makes the starter work more than necessary.

Anything that makes it more difficult to start the engine is bad for the starter. For instance, during low-ambient-temperature starts, use engine preheat and external power. A properly applied preheat helps the crankshaft turn more easily. Another problem associated with cold-weather starts is a weak battery. The application of external power assures full amperage to the starter immediately for quick engine cranking.

Running the starter for a long time is especially harmful to the starter motor. Most pilot operating handbooks publish starter time limitations, usually one minute or less. The limitation is related to cooling. Cranking the starter for longer than recommended causes the motor to overheat, which can melt the solder that holds the field and armature windings; the starter can literally self-destruct! If the engine does not start within the recommended starter limitation time, something probably is wrong with it. Don't continue to grind away; find out what the problem is with the engine. The three most common reasons an engine won't start are mixture in cutoff position, fuel selector off, or magnetos off.

Most airframe manufacturers recommend routine starter-system inspection and maintenance at least at the annual inspection. Some handbooks recommend inspection twice a year. The starter is susceptible to cumulative damage, and the manufacturer's recommendations should be followed closely.

Starter cable terminals should regularly be checked for corrosion. If present, wash the area with baking soda and water solution and dry. Then coat terminals with petroleum jelly. At the same time check for a loose terminal-to-battery stud connection and tighten it as necessary. Another vulnerable point is where the cables actually enter the battery terminal. They should fit tightly with no loose strands or partial breaks. Also inspect the general security and integrity of the insulation. These may seem like trivial points, but even loose strands can cause a loss of cranking power.

If there is concern about the system, you can conduct a voltage-loss test to locate high-resistance connections, which reduce efficiency. Using a low-reading voltmeter, have someone crank the engine while you check the voltage loss from battery post to start motor terminal. Maximum loss is .3 volts per 100 amps; a 200-amp system would allow a maximum voltage loss of .6 volts. The maximum voltage loss from the battery ground post to the starter frame is .1 volt per 100 amps. Great caution must be exercised when conducting this test. Be sure you disable the engine so it will not inadvertently start. The best way to do that is to remove the spark plug leads. Also, stay clear of the prop, as the starter can turn it fast enough for it to be lethal. Have an experienced mechanic do the test or at least check you out on how to do it safely.

TROUBLESHOOTING

The electrical system is probably one of the most difficult systems in an airplane to understand and troubleshoot. Any information you can give to your mechanic, even seemingly unrelated items, may make a difference. This is primarily due to the fact that many other systems somehow relate to the electrical system. For instance, a hydraulic gear system will use an electric pump. Another quirk of electrical appliances is that

Chapter Eight

some of them will continue to operate when there are problems with the system, but their accuracy will be off. Therefore, any tendency of electrical equipment to operate substandard should be brought to the attention of the mechanic.

Other occurrences of interest to the mechanic would be unexplainable dimming of lights, smoke, pungent odors that might indicate burning wire, and equipment that operates unpredictably. In general, the circumstances surrounding any problem will be of interest, such as the ammeter reading, other equipment in use, and conditions of flight.

One icicled, Illinois day a number of years ago, I attempted to start an airplane and absolutely nothing happened. After several attempts, I gave up and "informed" the mechanic that the battery was completely dead. He wanted to try it himself, but I was so certain about the diagnosis that he pulled the battery and brought it to the shop. Unfortunately, there was nothing wrong with the battery. It turned out the mag switch had worn out, probably from the giant key ring full of swinging keys that kept tugging at it during flight. I learned three important things that cold morning: Don't hang a bunch of keys from the mag switch. If the battery appears dead, you should check for excessive key movement in the mag switch, as it may be worn out. And perhaps the most important of all is let mechanics do their job.

Lead Acid Battery

Specific battery-related problems, for most electrical systems, are generally not too difficult for an experienced mechanic to readily solve, provided you do your part to help. You should explain the symptoms in detail, noting such things as how long the battery sat idle, what the ambient temperatures were during those days, and the conditions at the time of the problem. Table 8-8 provides a basic but helpful troubleshooting reference. It is also helpful if you can tell the mechanic any of the following.

- What the current draw has been according to the ammeter. If you have one of those that only indicates charge/discharge, then what did it indicate?
- Has there been a frequent need to replace water?
- Does the battery never seem to get fully charged?
- If you experienced difficult starts even in warm temperatures.
- If you attempted a start in very cold temperatures without pulling the prop through, getting a preheat, or using a ground power unit.
- Failure of specific electrical equipment.
- Pungent odors that might indicate burning wires.
- Unexplainable dimming of lights.

Answers to all these questions will help the mechanic to diagnose the problem. In general, the time-tested lead acid battery is reliable and trustworthy. A bad battery is more often the result of bad preventive maintenance. One of the first things a budding mechanic learns in school is the old saw, "If you take care of your tools, your tools will take care of you!"—a saying that applies equally well to the lead acid battery.

Table 8-8. Troubleshooting Batteries

Trouble	Cause	Remedy
Battery will not hold its charge	Battery life is beyond warranty.	Replace
	Charging rate set too low.	Check and correct the setting in accordance with instructions applying to regulating equipment.
	Discharge too great to replace.	Check battery for proper size and capacity. If too small or too low-rated capacity, replace with proper battery. Use of starter on ground and other electrical equipment in air must be reduced.
	Standing too long (hot climate).	Remove battery and recharge.
	Equipment left on accidentally.	Remove battery and recharge.
	Short circuit, or short to ground in wiring.	Check wiring and correct trouble, then recharge.
	Broken cell partition.	This is usually indicated by two or more adjacent cells running down continually. Replace battery.
Battery life is short	Overcharge.	This causes buckling of plates, shedding of active material, oxidation of grids, overheating, excessive loss of water. Check and correct adjustment in accordance with instructions applying to regulator equipment.
	Level of electrolyte is below top of plates	Keep electrolyte level above cell separators.
	Frequent discharges. This is due to excessive use of starter and other electrical equipment while on ground and recharging in air.	Reduce unnecessary use of starter and other electrical equipment while on the ground.
	Sulphated plates. This occurs when the battery is left in a discharged or uncharged (one-half or less) condition for a period of time, or electrolyte is not maintained at its proper level.	Charge at normal rate until the specific gravity does not rise for two hours and then give a 60 hour overcharge at 10% of the normal charging rate of the battery. If battery capacity is still low, replace battery.
Cracked cell jars:	Hold down loose.	Replace with fully charged battery.

Table 8-8. (Continued)

Trouble	Cause	Remedy
Cracked cell jars:	Frozen battery due to adding water in cold weather without charging the battery sufficiently afterward to thoroughly mix the water with electrolyte before letting it stand, or due to low specific gravity of the electrolyte caused by improper filling.	Replace with fully charged battery.
Compound on top of battery melts:	Charging rate is too high.	Check and correct setting in accordance with instructions applying to regulating equipment.
	Electrolyte on top of cells. Caused by overfilling. May short circuit the battery. The resulting heat will then soften compound.	Remove any electrolyte from top of neutralize with solution of sodium bicarbonate or ammonia. Then, wash battery thoroughly. Charge and replace in airplane.
Electrolyte runs out of vent plugs:	Too much water added to battery	Remove excess to correct electrolyte.
	Excessive charging rate.	Check and correct setting in accordance with instructions applying to regulating equipment.
Cell connector melted in center:	Shorted or grounded cable, causing direct full discharge of battery	Repair short or ground and replace battery.
Battery freezes:	Discharged battery.	Replace with fully charged battery.
	Water added and battery not charged immediately.	
Polarity reversed:	Battery connected backward on airplane.	Slowly discharge completely and then charge correctly and test before use.
	Battery connected backward on charger.	
Battery consumes excessive water:	Charging rate too high.	Check and correct setting in accordance with instructions applying to regulating equipment.
	Electrolyte runs out of vent plugs.	Level of electrolyte too high. Adjust.
Battery will not come up to charge:	Battery worn out.	Give capacity test and replace if capacity is low.
	Battery badly sulphated.	Charge as for sulphated.
	Improper storage. Dry batteries stored in a damp location, or wet batteries stored for too long a period without charging.	Charge as for sulphated plates.

NiCd Battery

Foaming during charge is an indication of low electrolyte concentration. Recondition the electrolyte and replace the old electrolyte in the foaming cells. Contamination of the electrolyte can be more serious because the contaminants may have caused permanent damage in the cells. This probably is going to require replacement of the affected individual cell.

If the battery output voltage is below normal, the most likely reason is you have accidentally left a load on the battery and it has discharged; simply recharge it. Other possibilities are simple, physical problems such as loose, dirty, damaged, or broken hardware; or loose, corroded, burned, or pitted connectors. Often a thorough cleaning of the problem area will rectify the situation. The worst case is a defective or reversed cell, which requires replacement of the cell. If the electrolyte level is too low, there is a cell imbalance, or the charging voltage has been too low; the battery should be reconditioned and the level of the electrolyte brought up. The voltage regulator also should be adjusted in the case of a low-charging voltage. No battery output could be the result of broken or disconnected hardware or links or a loose battery connector. Repair or replace the parts as necessary and recondition the battery.

Discolored, corroded, and/or burned hardware, connectors, or terminals indicate improper maintenance. The most likely problems are loose hardware, improper matching of parts, shorts between links, and improper cleaning. Correct the problem and recondition the battery if necessary.

If a cell is distorted or damaged, the problem could be a cell with an internal short, an overheated battery, or improper cooling. It also could be the result of charger failure or a plugged cell vent cap. Your mechanic will have to find the problem, correct it, and recondition the battery. Distortion of the battery case and/or cover could be the result of a major explosion of one of the cells, a dry cell, a high-charge voltage, a charge failure, or a plugged battery vent. Again, the mechanic will have to isolate and correct the problem and recondition the battery.

Starter

The operator may employ several troubleshooting techniques on the starter circuit. If the starter motor power seems insufficient to turn the engine, the most likely problem is the battery. If the battery is sound, inspect the load circuit; it could have a loose connection or a frayed cable. There also may be insufficient lubrication in the engine or something may be binding the engine or propeller. If the starter doesn't activate at all, the first thing to check is the battery and its connections. The starter switch also could be defective; try jiggling it.

If the starter draws unusually high current, there probably is an excessive load. Several things could be the culprit. If it's cold, the problem may be congealed oil. Use a preheat and external power for the next start attempt. There also may be an obstruction to the propeller or, in the worst case, the engine bearings may have seized.

Occasionally, a starter will run too fast. This almost certainly is caused by excessive external power output voltage. Should the starter overheat, frequently made obvious by its odor, there could be several causes. Most often it is caused by exceeding recommended starter time limitations. Bad engine bearings will cause the engine to turn hard, which will overload and overheat the starter. Also, excessive voltage can overheat the motor. Whenever you smell an electrical problem, it is prudent to stop whatever you are doing and investigate. Similarly, while aircraft tend to be noisy, excessive starter vibration or noise should be investigated. The probable cause is a loose or broken mounting. I can think of few situations less desirable than having a starter motor flying around loose inside an engine cowling during flight.

9
Propellers

WITHOUT PROPELLERS, THERE WOULD BE NO AIRPLANES TODAY. IT IS the propeller that first made powered flight possible, and it is the propeller that caused the first powered flight fatality. Lt. Thomas Selfridge, who was evaluating the airplane for the Army, was killed while flying with Orville Wright on the Wright Military Flyer. The accident, which occurred on September 17th, 1908, was the result of a prop failure.

While propellers have come a long way since 1908, they can still be a major problem. The Worldwide Aircraft Propeller Association did a random survey of 15 U.S. propeller repair stations, approximately 12% of the total. It showed that approximately 62% of the 2743 propellers submitted for inspection were unairworthy due to corrosion alone. The care and feeding of propellers is something with which every pilot and aircraft owner should be familiar.

SYSTEM OVERVIEW

A propeller can be defined as a rotating airfoil with two or more blades attached to, or integral with, a hub. Powered by an aircraft engine, its purpose is to convert engine horsepower to thrust. Propellers are designed to be compatible with specific engines on specific aircraft and are not interchangeable, except as specifically permitted by the manufacturer.

There are two basic configurations for propellers: tractor and pusher. The tractor propeller, mounted on a forward-facing engine, is most common. It's so named because it pulls the airplane along like a "tractor." The pusher prop is mounted on a rear-facing engine and pushes the airplane from behind. One twin-engine aircraft, the Cessna 337 Skymaster, uses a tractor prop on the nose of the fuselage and a pusher prop on the back. The aircraft has twin-engine performance with the safety and ease afforded by centerline thrust, which does not present an asymmetric thrust controllability problem should one engine fail.

While most light aircraft have two-bladed propellers, three or more blades are very common on high-performance singles and multiengine aircraft. There are several advantages to having more than two blades. The individual blades can often be shorter, allowing increased ground clearance, without a decrease in performance. Shorter blades have higher, and less objectionable, sound frequencies and an overall reduction in vibration. The additional blades also produce a greater flywheel effect and generally improve aircraft performance at takeoff and climb speeds.

Propeller tips may be rounded or squared, depending on noise requirements, blade-vibration characteristics, and other special design considerations. Tests have shown that elliptical tips are slightly more efficient than square ones, but that's not the reason square tips are more common; they're more practical. Square tips leave extra material, which can be removed after damage occurs, turning them into round or elliptical tips, for instance, and still maintain the required prop diameter. That's important because prop diameter (the circle circumscribed by the blade tips) is carefully designed for maximum efficiency. Slower aircraft will use a larger diameter while high performance aircraft will have smaller diameter props. The longer the propeller blade, the greater the distance the tip must travel during each revolution. High-speed aircraft engines can cause the tips of long propeller blades to reach the speed of sound. This results in significant aerodynamic breakdown, vibration, and reduced performance.

Terminology

Two terms that frequently confuse pilots are prop face and camber. A look at Figure 9-1 will help clear up some propeller terminology. The flat side of the propeller, which faces the pilot, is called the face; the opposite, cambered side is, logically enough, called the back or camber side.

Blade angle is the angle between the plane of rotation and the chord line of a given propeller airfoil section. Blade angle decreases along the length of the blade (called pitch distribution) with the greatest angle nearest the hub. The reason a prop has pitch distribution is because the tip must travel at a higher relative velocity than the hub. To see an example of this, tie a key to a length of string and spin it over your head. The key and string, essentially one continuous object, must turn at the same RPM (revolutions per minute), but the key covers significantly more distance. This is where the prop most significantly differs with the wing, which is an airfoil with a fairly constant relative wind throughout its span. Pitch distribution attempts to make the prop efficient along its entire length to obtain maximum forward movement. The distance a prop section moves forward in one revolution is called pitch and is measured in inches. The amount of lift produced by any given prop will vary with airfoil shape, angle of attack on blade sections, and prop RPM.

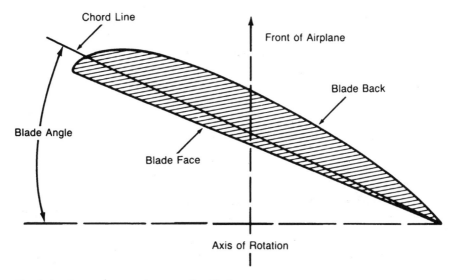

Fig. 9-1. *Cross-section of a propeller blade.*

Pitch distribution makes talking about a specific blade angle difficult because the angle gradually changes through the entire length of the prop. So what we need is a kind of "street address" to help us identify specific blade angle locations. Blade station is a specified distance, measured from the center of the hub outward. For instance, in one model of the Beech B-55 Baron, the Pilot's Operating Handbook (POH) describes a feathered condition as one where there is an 80.0-degree blade angle at the 30-inch station.

CONTROLLING PROPELLER PITCH

Over the years, several different types of propellers have been developed. The ground-adjustable type, found on older aircraft, permitted the pilot to set prop pitch on the ground. Normally set to provide maximum efficiency during straight-and-level flight, it could be reset to an angle more conducive to short-field takeoffs; the penalty would be slower cruise performance. The problem was that most fields were short in those days, so cruise efficiency suffered; a two-position prop was developed. Now, from within the cockpit, the pilot could switch from low pitch for takeoff to high pitch for cruise. The next logical step was a controllable pitch propeller, which allowed the pilot to select any blade angle within the prop's range, but it was a matter of guesswork to decide which was the best. The short-lived automatic pitch prop had the capability of setting its own pitch as a result of aerodynamic forces; the pilot had no control at all!

FIXED-PITCH PROPELLERS

The fixed-pitch propeller is as common today in light, single-engine trainer aircraft as it was during the heyday of the Wright Brothers. The wooden version of this reliable one-piece prop, long a coveted decoration for the pilot's den, is still providing excellent service in the air, though it is rarely, if ever, used in the manufacture of new aircraft.

Construction of a wooden prop, as depicted in Figure 9-2, consists of several laminated layers of wood with a doped cotton fabric sheathing glued to the last 12–15 inches of the blade. Occasionally a plastic coating will be used instead of fabric, but they both serve to reinforce the thin blade tip. On top of this runs a metal (usually brass, Monel, or stainless steel) tipping along the leading edge of the prop, which helps prevent foreign object damage. Small holes are drilled into the metal tip to allow the wood to breathe and moisture to escape.

For ease of maintenance and reduced weight, most modern fixed-pitch props are made of aluminum. Fixed-pitch props can be purchased, depending on the operator's needs, for maximum efficiency during climb or cruise. Most aircraft would have the latter installed, but aircraft primarily used for hauling skydivers should have a climb prop. It makes the climb to altitude much quicker, sometimes even before the first-time jumpers lose their nerve. Other operations that benefit from a climb prop are aircraft that operate routinely off of a short, sod strip, or those that tow gliders.

CONSTANT-SPEED PROPELLER

The modern constant-speed propeller allows the pilot to select an engine RPM based on current operating conditions. The prop governor adjusts the blade angle to maintain selected RPM. This type of prop is used on most medium- and high-performance singles and practically all propeller-driven multiengine aircraft. One significant advantage is the ability to reduce prop drag to near zero (called feathering) should an engine fail in flight. When feathered, the prop blade turns its edge into the wind and the prop comes to a stop. Some aircraft, typically large recips and turboprops, even have the ability to rotate the blade angle to a negative value, which effectively creates thrust in the opposite or reverse direction. Prop reverse significantly reduces landing roll, and when combined with differential power (varying left and right engine power to aid steering) it greatly improves ground handling.

A constant-speed propeller system is one in which propeller blade angle is varied by a governor so that a constant propeller RPM can be maintained despite engine throttle changes or variations in aircraft speed. This allows the pilot to operate the propeller at a high level of efficiency. Unlike the fixed-pitch propeller, which is efficient under only one set of circumstances (either in cruise or climb, depending on type of propeller), the constant-speed prop permits both efficient climb and cruise. Figure 9-3 compares the takeoff climb performance of both types of propellers. For any given phase of flight, there is one RPM setting that is most efficient; it gives you the greatest gain for the least amount of energy. The constant-speed prop allows you to operate, within limitations, at that RPM.

For a propeller to turn at a constant RPM as power conditions change, the angle of attack of the blades must be adjustable. Because it would be impractical for the pilot to make such frequent adjustments, it is done automatically by the propeller's constant-speed governor.

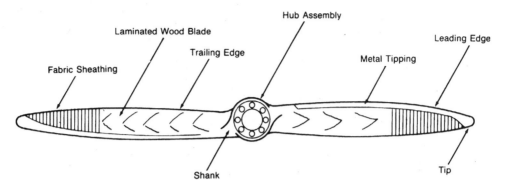

Fig. 9-2. *Parts of a wooden propeller.*

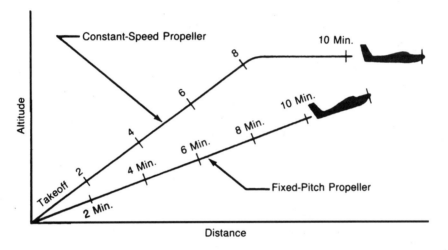

Fig. 9-3. *Climb comparison.*

Controlling Blade Angle

Fixed force and variable force are the two opposing forces that control blade angle. During operation, a fixed force either tends to increase or decrease blade angle, depending on the particular design. This fixed force can be caused by centrifugal force acting on counterweights, by a spring, or simply by centrifugal twisting moment. Variable force, which causes the blade angle to change by counteracting the fixed force, is actuated by a governor that controls oil pressure on a piston in the propeller dome. The piston is connected to the blades through a mechanical linkage; this linkage converts the linear, hydraulic motion into a rotary motion necessary for changing the blade angle.

In certain McCauley propellers, for instance, oil pressure on a piston is used to increase blade angle. When the governor diverts oil away from the piston, centrifugal twisting moment on the blades and a booster spring in the prop hub cause the blade pitch to decrease. Other systems work just the opposite, but the principle is the same.

Chapter Nine

The governor, connected by a driveshaft to the engine drivetrain, senses engine RPM and compares it to the RPM the pilot has selected with the propeller control lever. It redirects oil pressure to, or from, the propeller dome, which changes the blade angle to maintain that RPM, as shown in Figure 9-4. An oil pump drive gear located on the driveshaft meshes with an oil pump idler gear and squeezes oil outward, boosting the engine oil pressure to that required for the propeller system. A pressure relief valve bypasses excess pressure to the inlet side of the pump. Oil goes through the hollow driveshaft to the

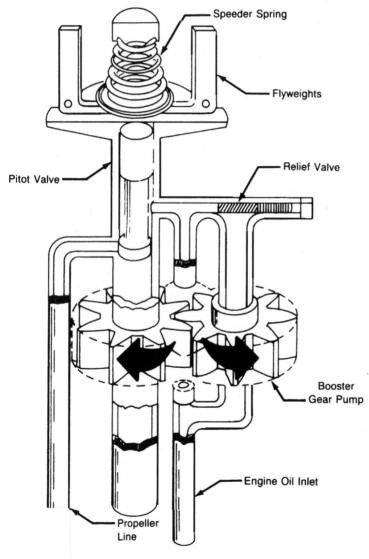

sic governor configuration.

pilot valve on the governor, which moves up or down inside the driveshaft to direct the oil through different ports. One port diverts oil pressure to the propeller dome; another allows it to flow back, relieving the pressure.

Pilot valve position is controlled by a set of flyweights mounted at the end of the driveshaft. The flyweights, sensitive to centrifugal force, tilt outward as RPM increases, inward as it decreases. In the outward position, the flyweights raise the pilot valve; in the inward position, they lower it. Pressing down on the pilot valve is a speeder spring, which is connected by control cable, pulley, and speeder rack to the cockpit propeller lever.

If the pilot wants a higher engine RPM, the propeller control lever is pushed forward. This compresses the speeder spring, which in turn pushes down on the flyweights, tilting them inward. The propeller, which has been turning at a lower RPM than is now desired, is in an "underspeed" condition. As the flyweights are tilted inward, they move the pilot valve down, permitting oil to flow under pressure to the propeller dome. The oil pushes on the piston, which in turn decreases the blade angle. The lower blade angle allows the propeller to turn faster under the given conditions, so RPM increases. When it does, centrifugal force on the flyweights slowly overcomes speeder spring force and the pilot valve returns to the neutral position. This stops the oil flow and maintains a constant blade angle.

If the pilot wants a lower RPM, the propeller control is moved aft, which relaxes speeder-spring tension, tilting the flyweights outward. This condition, called "overspeed," occurs whenever existing RPM is higher than that selected with the propeller lever. The result is the flyweights raise the pilot valve, permitting the oil to flow out of the propeller dome, causing the blade angle to increase, and thereby causing RPM to decrease. Decreased RPM results in decreased centrifugal force on the flyweights; slowly, the flyweights again succumb to speeder-spring force. The pilot valve returns to the neutral position, the oil flow stops, and blade angle remains constant until the next disturbance. If this sounds involved and complicated, then consider that it occurs almost instantaneously. Overspeed and underspeed conditions are corrected so quickly, the pilot typically is unaware they have occurred; they don't even show up on the tachometer!

Of course, there are limitations to this process. The governor can maintain a given RPM as long as the blade angle remains between the low- and high-pitch stops, but it also requires horsepower. For instance, every airplane can outclimb its maximum sea-level RPM if allowed to go high enough. Normally aspirated aircraft will do so quickly, in fact. But there is another source of horsepower available to the propeller besides the engine: free-stream energy. The governor doesn't care if the engine turns the propeller or the propeller is turning the engine as long as it does it at the selected RPM! Placing the engine at idle and putting the aircraft in a fast descent can cause the airflow to turn the propeller, which in turn can cause the propeller to turn the engine. If the prop is set for 2200 RPM, the governor will try to maintain that RPM, regardless of throttle setting, by adjusting the blade angle as necessary.

Taken to the extreme the governor eventually will run out of blade travel. When that happens the prop functionally becomes "fixed pitch," with RPM directly affected by throttle setting (manifold pressure). But a windmilling constant-speed prop has sufficient torque at cruise to maintain the RPM on a dead engine! That is why it is virtually impossible to identify visually which of two or more reciprocating engines have failed in cruise flight. The windmilling propeller will turn the engine and its accessories—the generator, fuel pump, vacuum pump, air conditioner, everything. To the pilot there is no engine-related instrument indication of engine failure! It is nonetheless obvious because there is an incredible increase in drag on the side of the inoperative engine requiring the pilot to use heavy opposite rudder to keep the aircraft from yawing into the dead engine.

One student of mine suggested that there was an engine-related instrument that would indicate engine failure—the cylinder head temperature gauge (CHT). His theory was that the CHT would indicate that the failed engine was cooling down. Fundamentally that is correct, but there is a considerable time lag involved. If the pilot can't figure out which is the failed engine before it cools down enough to register on the CHT gauge, it is more likely they'll be determining the failed engine in the accident investigation afterwards!

Propeller Feathering

Engine failure in a single-engine aircraft always means a landing, but with two or more engines that is not necessarily the case. To make the airplane flyable on one engine, it is necessary to reduce drag to a minimum. The solution is to turn the edge of the windmilling propeller blade into the wind. The procedure, called *feathering*, sets the average blade angle to approximately 90 degrees, as depicted in Figure 9-5, which stops the propeller in flight. The pilot may feather a prop by pulling the appropriate lever all the way back to the low RPM setting, through the safety detent, and into the full aft, feathered position. Some aircraft, particularly large, older ones such as the DC-3, have a pushbutton control rather than prop-lever detent, but the result is the same. Feathering is totally independent of, and overrides, the constant-speed operation. In some aircraft, the propeller doesn't even have to be turning to be feathered.

The McCauley feathering system, common on many Cessna and Beech twins, uses counterweights and an internal spring that overrides oil pressure to feather the prop. That provides an inherent safety feature: If you lose oil pressure, the propeller automatically feathers. Not a bad idea, considering that an engine without oil will freeze up! A spring-loaded latch automatically engages during normal engine shutdown to prevent the prop from feathering.

When you feather a prop for practice, there are two ways to unfeather it in flight. Some aircraft, especially those used primarily for training, have unfeathering accumulators. This cylinder contains a diaphragm or piston that separates an air (or nitrogen) precharge from system oil. The pilot moves the prop lever out of feather to the full-forward position. When the unfeathering accumulator is activated, the high-pressure oil charge is moved to the propeller dome cylinder and forces the blades to a lower blade angle. Once the blade angle has been changed, the force of the relative wind begins to turn the prop with a dramatic increase in drag. At that point, fuel may be introduced and the magnetos turned on; the engine should start normally. Remember that the engine oil and cylinders will have cooled; give them time to warm up at a low power setting before returning to cruise power.

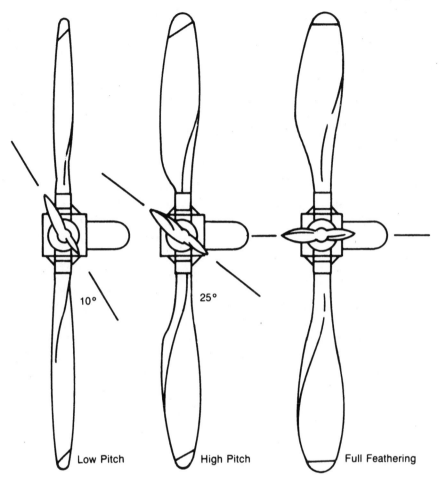

10°

25°

Low Pitch

High Pitch

Full Feathering

Fig. 9-5. *Blade angles.*

An aircraft without an accumulator requires that the pilot put the prop lever full forward and get the engine turning by using the starter. As the engine moves the oil, the governor builds up oil pressure, which overcomes the force of the spring and moves the blade to a smaller angle. While this sounds easy, there are a few potential problems.

One problem is that it's more work for the starter; but more significant, the aircraft is totally reliant on the electrical system and starter. If any problem develops with either of those systems once the prop is feathered, you may have a real emergency on your hands. Hartzell feathering systems are used on Piper, Aero Commanders, some Beech, and older Cessnas. While there are differences, they are fundamentally similar to McCauley in both the feathering and unfeathering modes.

Preflight procedures from the pilot's operating handbook (POH) to test feathering capability should be followed closely. It is important to exercise the prop during preflight, especially on cold days when the oil tends to congeal. To do this, run the engine up to the recommended manifold pressure with the prop control in its full forward position. Then quickly pull the prop lever all the way back to the feather detent, get about a 500-RPM drop, and then push it forward again. The entire action should be fairly smooth: pull, drop, push. If the oil is cold, you may not get an RPM drop immediately, so push the prop lever back and forth fairly rapidly to get warmer engine oil moving through the lines. Once you begin to see a response in RPM, continue with the normal prop feathercheck. Most manufacturers recommend that you periodically do a complete feather procedure just to make sure everything will work should you need it; consult your POH.

Propeller Control

Fixed-pitch props are directly controlled with the throttle; the reference instrument is the tachometer, which indicates engine revolutions per minute. A constant-speed propeller actually is controlled by a propeller lever in conjunction with the tachometer. The throttle controls engine manifold pressure and is referenced by the manifold pressure (m.p.) gauge, which indicates inches of mercury. With the constant-speed system, setting power correctly is essential to efficient operation and even engine life. A pilot always should consult the POH for specific combinations of RPM and m.p. for various climb and cruise configurations. Excessive m.p. for a given RPM could cause engine damage. For instance, an m.p. too high for the RPM will cause abnormal cylinder pressure, leading to high cylinder-head temperatures, detonation, and excessive stress on engine parts. A good rule of thumb is when increasing power, increase RPM first, then m.p.; to decrease power, pull the throttle back to approximately 1 inch below the desired m.p., then pull back the prop lever to the new RPM setting. The reason for the 1-inch difference is because reducing RPM will cause the m.p. to rise slightly.

TROUBLESHOOTING

There are significant differences between constant-speed propeller systems; always consult your POH for proper troubleshooting procedures. In general, all nonfeathering propellers will go to high RPM if there is a loss of oil pressure; the feathering types will go to feather in the same situation. If there is damage to the prop-lever linkage, the governor is spring-loaded to high RPM. The reasoning is, if you don't have feather, then the propeller should go into the condition most conducive for a go-around—low blade angle (high RPM). Failure to unfeather in flight when you don't have accumulators probably means there is insufficient starter power to turn the engine enough to build up oil pressure in the governor. This is not an uncommon problem, particularly in older aircraft.

Before attempting a single-engine landing, you might want to try placing the aircraft in a shallow dive and momentarily activating the starter, a procedure called "bumping." The increased airspeed combined with the engine rotation from the starter might turn the propeller enough to get the oil moving and start the unfeathering process. If you have an accumulator and the prop fails to unfeather, it probably is the result of insufficient or no air charge, an oil leak in the accumulator or hoses, or a ruptured diaphragm, permitting the air charge to leak into the oil. In any case, there isn't anything you can do about it in flight. You will have to revert to using the nonaccumulator unfeathering procedure for your aircraft.

Hartzell feathering propellers with counterweights or spring-assist use an air charge to help feather the prop. Low air charge is indicated during the preflight feathercheck and in flight by sluggish or slow RPM control, especially when reducing RPM. There also may be some difficulty in maintaining RPM, but before you curse the system, make sure you don't have creeping throttle or prop levers. Over the years I have watched countless students torque the control lever's friction locks as tight as they would go, then muscle the levers back and forth. Then, after the friction locks are completely worn out, they can't figure out why the controls creep by themselves.

Other indications of an air charge problem would be minor overspeed problems, particularly with rapid throttle application, poor RPM recovery under the same situation, and poor synchronization in the upper cruise speed range. If you are in flight when the problem occurs, reduce the throttle and airspeed until RPM control is regained. Be careful not to go below best single-engine, rate-of-climb speed for your aircraft. Once you have regained control, increase the throttle slowly to get as much power back as possible without returning to an overspeed condition.

Noncounterweighted models work essentially like McCauley systems. Oil pressure increases blade angle; centrifugal twisting movement on the blades decreases it. Counterweighted models use oil pressure to decrease blade angle and centrifugal force on the counterweights to increase it. Be particularly careful to watch for grease leaks near the propeller hub. Common causes are loose, missing, or defective zerk fittings, defective or loose blade-clamp seals, and overlubrication of blades. Have a mechanic check out any leaks.

The Compact Propeller version with a low air charge would experience improper constant-speed operation, overspeed, and a surge tendency. An excessive air charge would cause an inability to achieve maximum RPM and could cause the propeller to feather on the ground when the engine is shut down normally.

PREFLIGHT AND RUNUP

For all types of constant-speed propellers, check during preflight for oil and grease leaks near the blade shanks. Inspect the spinner for security. The spinner is typically not an optional item and must be in place for proper cooling airflow to occur. Look for excessive looseness of the blades, but realize that some play, called *blade shake*, is inherent in the design. Whenever working on or near the propeller, be certain the magnetos and master switch are off, the mixture is in cutoff, and avoid getting into the propeller arc. Always follow the POH preflight procedures. Should you discover your prop is suffering from reduced or lost air charge, use very gentle throttle movements to prevent potentially damaging prop-overspeed conditions.

Good procedures dictate that the pilot avoid runups on areas of gravel, stones, broken asphalt, or loose sand, for two reasons. First, debris in the runup area will cause prop blade erosion. Second, the power of the propwash will pick up loose debris and throw it back onto whatever is behind you. It is prudent to avoid any operations on gravel runways, but if you have no choice, use the following takeoff procedure, provided runway length permits: let the aircraft begin rolling while at low throttle, then gradually increase to takeoff power. This procedure will reduce the possibility of the prop picking up a stone and nicking the blades.

In general, it is a good idea to taxi slowly and cautiously to minimize hazards such as foreign objects on runways, snowdrifts, taxiway and runway lights, and tiedown chains. Always avoid pulling or pushing the aircraft by its propeller, which for some inexplicable reason seems to be a passion of flight instructors. Pulling an airplane by the prop tip, which is admittedly uncommon, creates a high probability of bending the blade. Pull an airplane around by the prop down near its hub, the mistake of choice, and you can stress the crankshaft.

If it is absolutely necessary to move an airplane by the prop, grab the propeller shank as close to the hub as possible and slowly increase pressure until the airplane begins to move. But remember that one damaged propeller will cost many times the price of a towbar. Also, be very careful to avoid putting a rotational force on the prop, as there is the ever-present threat of a hot mag starting the engine. Always assume the airplane has an ungrounded magneto, and any rotation to the propeller will cause the engine to start. I have personally been on two accident investigations, one regarding a lost limb and the other a lost life, due to props that roared to life unexpectedly in exactly that manner.

The dangers associated with a static aircraft and a turning propeller cannot be overemphasized. At the very best a turning propeller is a life-threatening environment. Take great pains to avoid loading and unloading passengers with a turning prop. If you're dropping off or picking up passengers in a twin-engine aircraft and for some reason you can't completely shut down, you want to position the aircraft so the door faces the ramp gate. Then shut down the engine on the side of the door. As a rule, never start an engine when there are people in the vicinity of the airplane, and be especially wary of children, animals, and machinery.

PROPELLER MAINTENANCE

The prop, considered a separate component of the airplane, typically has a manufacturer recommended time between overhaul (TBO) different than the engine. Know its TBO and follow the manufacturer's recommendations. During each preflight, the pilot should check for wear, nicks, dents, and other damage. If a problem occurs, it either should be dressed by an A&P or referred to a propeller shop if the damage is too great. Every 100-hour or annual inspection, whichever comes first, the prop should be carefully inspected by an A&P who also will remove the spinner, check the hub parts for wear and damage, check the air charge, and lubricate it as required. It also is a good idea to have the tachometer checked for accuracy periodically. If your tachometer has an RPM red arc, you should have your tach checked every 100 hours for accuracy to make sure you're not inadvertently operating in the red arc area!

Propeller Balancing

Under the best of circumstances, props go out of balance with age, primarily from blade erosion. An out-of-balance prop results in vibration, which leads to the premature failure of such parts as the alternator, fuel control, engine wiring harness, and avionics. It also causes oil cooler leaks; broken or cracked engine mounts, exhaust manifolds, and turbocharger mounts; and sheet metal cracks in the fuselage and cowlings. There are two approaches to balancing a propeller: static and dynamic.

To static balance a prop, it must be removed from the airplane and put on a stand in a shop. The problem with this method is it doesn't take into account the effect of the engine, its accessories, bulkhead, and prop spinners. Dynamic balancing is accomplished while the propeller is on the airplane, taking all aerodynamic forces into account. Also, it can be done at different RPM settings. Many shops now use the Chadwick-Helmuth Vibrex Dynamic Balancer. With this system, nothing is removed from the propeller or the engine; it uses a small velocimeter to measure movement produced by the out-of-balance condition at different RPMs. The instrument then tells the mechanic how much weight must be added to the hub to balance the prop; the process is analogous to dynamic balancing of tires on a car.

It is recommended that a prop be balanced every time it, or the engine, is overhauled. Some manufacturers also suggest a prop should be balanced at 500-hour intervals. It's also a good idea to have it balanced whenever you have a cylinder reworked, after any significant repairs to the prop or engine, and at the onset of any unusual vibration. The benefits are immediate and noticeable. The vibration disappears and you get the long-term benefit of extended component life.

PROPELLERS AND THE FARS

The manufacture of propellers is very tightly controlled by Federal Aviation Regulations (FARs). They not only define the limitations of the prop but also its effect on the engine. The single engine prop must limit engine RPM to the maximum allowable when the engine is at full power and the aircraft is at its best rate of climb speed. The purpose of this requirement is to prevent engine damage due to overspeed. It is for this reason that during maximum-power ground runup in a zero wind condition, you are unable to reach redline on the tachometer. In addition, the prop must prevent the engine from exceeding rated RPM by no more than 10 percent in a closed throttle dive at the aircraft's "never exceed" speed. The constant speed prop must always restrict the engine to rated RPM during normal operations. In the event of governor failure, static RPM must not exceed 103 percent of the rated RPM. This is essentially what determines where the manufacturer sets the low blade (high RPM) angle.

Similarly, after years of practically no standardization, the FARs now detail the design of cockpit controls and instruments. Forward movement of controls produces an increasing effect so mixtures enrichen, prop RPM gets higher, and forward thrust increases. An aft movement of the throttle on an aircraft with thrust reverse will place the prop blade at a negative angle and increase the reverse thrust. It is also required that controls be easily distinguished from one another as to shape and color. Instrumentation must use standardized markings: a red radial line indicates maximum operating limitation, a green arc is the normal operating range, a yellow arc is the takeoff and precautionary range, and a red arc is the critical vibration range.

Another critical area covered by regulation is minimum terrain clearance or the space between level ground and the edge of the prop tip. Ground clearance for land aircraft assumes normal inflation of struts and tires. For conventional-gear (tailwheel) aircraft, it must be at least 9 inches in the takeoff attitude; for tricycle gear it is 7 inches in the most nose-low, normal attitude, whether stationary, during taxi, or takeoff. It also requires positive clearance with strut and tire deflated. For the seaplane, minimum prop-to-water clearance is 18 inches.

PREVENTIVE MAINTENANCE

Probably the best preventive maintenance possible is to keep the props clean. If a prop is dirty it is very difficult to see cracks and other problems while they are still repairable. To wash wooden props, use a soft brush or cloth and apply warm water and a mild soap across the entire blade. When finished, dry with a soft towel. Metal props should be cleaned with a non-oil-based cleaning solvent approved by the manufacturer, such as Stoddard solvent. Never use a caustic cleaner or acid. The solvent may be applied with a soft brush or cloth, then wiped dry with a soft cloth. After drying, wax the blades with a good auto paste wax to protect them from corrosion. For the same reason, it is also a good idea to wipe the blades periodically, even after every flight, with a damp cloth.

Propeller repairs typically have to be accomplished by the manufacturer or an authorized repair station. Some may be done by a powerplant mechanic, but there are no repairs authorized for the non-A&P. The owner/operator is primarily limited to basic preventive maintenance, which is essentially no more sophisticated than replacing defective safety wire or cotter keys.

It is permissible to lubricate parts of a propeller not requiring disassembly beyond nonstructural fairings, coverplates, and cowlings. You may also apply coatings such as paint, wax, and other preservatives if they are not prohibited by the manufacturer, are not contrary to good maintenance practice, and only if no disassembly is required. Major alterations and repairs may only be performed by an authorized repair station according to Advisory Circular (AC) 43.13-1, Acceptable Methods, Techniques and Practices. This would include repairs to deep dents, scars, cuts, and nicks.

Prop repair tolerances for cuts and dents are tight with little room to work. For instance, take a prop with a diameter of less than 10 feet, 6 inches. A nick located between the hub and the 24-inch station would not be allowed to exceed ³⁄₆₄ths of the blade width. If the blade width was 6 inches, a nick deeper than approximately ¼ inch would be cause for a repair facility to scrap the blade! If that seems unreasonable, consider the forces acting on the propeller.

Vibration is a natural by-product of the propeller producing thrust, but excess vibration will fatigue the metal and lead to failure. Mechanical vibration, the result of piston engine power pulses and crankshaft resonance, is a prime culprit in metal fatigue and prop failure. Some RPMs are particularly harmful to the prop, so the manufacturer puts a red arc on the tachometer. Operation is not permissible in the red arc except to pass through it, which is why it is important to have the tachometer checked for accuracy every 100 hours.

Thrust, aerodynamic twisting, torque, and centrifugal force (as much as 20 tons) are operational forces that act upon the propeller. With so much working against the propeller, there are amazingly few failures. Generally, prop failures are caused by fatigue cracks resulting from nicks and other scars left unattended. Therefore, because of the potential for problems more so than the likelihood, manufacturer-prescribed routine maintenance should be carefully adhered to. This is especially true for manufacturer-recommended overhaul of the constant speed prop. Typically based on hours in service and calendar time, it includes complete disassembly, inspection, reconditioning, and replacement of parts as necessary, and reassembly.

Despite years of faithful service and their aesthetic value, the wooden prop does have a wide range of potential problems. In particular, delamination (separation of laminations) is cause for hanging it in the den; repair is possible if only the outer lamination has begun to separate. Dents and other scars indicate real trouble in wood because they indicate cracks, and cracks lead to failure. If caught in time, small ones may be repaired with an inlay. Other minor defects may be curable with filler; small cracks, parallel with the grain, may be stopped with resin glue. Tip fabric should always lay neatly and metal tipping should be smooth and in good repair, or a repair station should be consulted as these problems tend to worsen rapidly. Cracks in the solder joints near the metal tipping may indicate wood deterioration beneath and always merit a thorough inspection by a mechanic. Some defects are beyond repair. They include a crack or cut across the grain, elongated bolt holes, warped blades, nicks or chips with significant wood missing, oversized crankshaft bore, a split blade, and cracks between the bolt-attach holes.

Aluminum alloy propellers have many advantages over their wooden equivalents. They are thinner with equal or greater strength and often weigh less if they are of the one-piece construction type. Leading and trailing edge defects may be dressed out by a power-plant mechanic, provided the finished size is less than 1/8th inch deep and 1.5 inches long and the repair has smooth and gradual curves. A slightly bent blade can be repaired, but there are precious few degrees of freedom.

Corrosion

The snake-in-the-grass for aluminum is corrosion; owners should do all they can to ensure against its insidious effects. Corrosion forms tiny, deep cavities that may extend inward, tunneling under the surface of the propeller and reappearing elsewhere—chemical wormholes eating your airplane. Props with de-ice boots and leading edge abrasion boots should be carefully checked for corrosion during 100-hour inspections because the damage may be hidden. Corrosion is especially a problem for aircraft routinely operating in high moisture or saltwater areas, and these aircraft should receive more care. Never attempt to remove corrosion with steel wool, emery cloth, steel (except stainless steel), wire brushes, or severe abrasive materials. Particles of steel wool or emery cloth will become embedded in the aluminum and lead to an even greater corrosion problem.

Chapter Nine

PREFLIGHT

Preflight of the propeller should always begin with checking to make sure the mags and master switches are OFF. Even so, always stay out of the prop arc during your preflight and never lean over the prop. After visually inspecting for corrosion, while standing clear, grasp the blade tip and test for looseness by pulling fore and aft. Some movement, called *blade shake*, is normal for constant-speed propellers because of their design. Run your hand over the face and back of the prop and your fingernail over the leading and trailing edges searching for irregularities of any kind. Nicks in the blade radiate lines of force outward, causing irreversible damage. If small and caught early enough, they may be dressed by a mechanic. If a small nick occurs while the airplane is in a location where no mechanic is available, as an emergency measure only, you might very lightly dress it with a fine, half-round file, just enough to smooth the jagged edge. Do not use emery cloth because particles will become embedded in the soft aluminum material and lead to corrosion. Then fly the airplane to the nearest A&P. If in doubt about its severity, don't fly it. Blade failures typically occur within inches of the prop tip, but there have been cases of failure as far inboard as the hub, so don't be fooled.

Contrary to popular belief, most prop spinners are not optional items, though it is not uncommon to see them inexplicably missing on aircraft. They are typically used to assure smooth cooling airflow into the cowling. Spinners should be checked for cracks, security of attachment, and evidence of oil. Oil indicates crankshaft seal failure. While standing near one end of the prop, sight down one tip across the blades to the other tip and check for alignment. If there appears to be a bend, you should perform this simple blade track check, which assures one blade tip follows the other in the same plane.

After double-checking that mags and master are OFF, point one prop tip directly down. Block up a smooth board directly under the prop tip and pull gently on the tip. Mark the spot on the board with a pencil, push the tip gently backward, and mark that spot with the pencil. Without moving the board or the airplane, rotate the other prop into the down position and repeat the procedure. The two sets of marks should not vary by more than ⅟₁₆th inch. If they do, have it checked by a mechanic.

Safe operation of propellers is mostly a matter of common sense. Without a doubt the prop is potentially the most lethal part of an aircraft, even when the engine is not running! Connected directly to a reciprocating engine, moving the prop means turning the engine over. That includes turning the mags and pulling air through the carburetor. Any number of system-related failures, ranging from a broken magneto P-lead to a fuel leak, could cause the engine to start unexpectedly. Turning a prop by hand, even backwards, should always be considered unsafe and avoided if possible.

The old routine of putting a prop in the vertical position to indicate a refueling order is an accident looking for a place to happen; use a windshield sign instead. And again, pulling an airplane around by its propeller is not a good idea. Use a towbar; it's much less expensive and time-consuming than an airplane out of control on the ramp running over your unattached arm. Another worthwhile rule of thumb is never get near a prop with a nonpilot in the airplane; there have been too many instances where a passenger has accidentally turned on the mags, pushed a mixture lever forward, or even hit a starter.

There are few nevers in aviation, but an engine should never be started when there is an unattended small child in the area. In general, it's really a good idea to avoid starting whenever there are people nearby. Always shout "clear," "prop," or your favorite term immediately prior to actual engine start, regardless of where the checklist puts it. Then look around the aircraft before engaging the starter; to the uninitiated, the word "clear" makes them look up to the sky and smile knowingly.

Even with all these precautions, some people can be mesmerized by a rotating propeller. Some years ago a pilot's wife walked back to the airplane for one last goodbye. She walked right through the running prop and lost an arm. To help reduce the potential for blindly walking into a prop, it is a good idea to paint the tip with alternate-colored stripes such as red tip, yellow stripe, and then red stripe. The face of the blade, which faces the pilot, should be painted flat black to prevent the pilot from being subjected to strobe effect. This problem occurs when flying with the sun at your back or at night with the aircraft lights on.

If you have reason to believe there is an impending blade failure while in flight, reduce the power to idle. The slower RPM will help you observe the prop to assess the situation. If there is an actual separation, immediately shut down the engine, as the resultant vibration could easily tear the engine from its mounts. While the propeller is a highly reliable piece of equipment, failures do occur.

10
Landing Gear Systems

OF ALL THE SYSTEMS ON AN AIRCRAFT, LANDING GEAR CAN certainly be the most embarrassing. It has the ability to quickly cause extensive damage by the pilot simply forgetting to move a single switch. And never be lulled into thinking the landing gear warning horn will prevent you from making that costly mistake.

A military pilot—who successfully got his single-engine fighter back to an airport after the engine had failed some distance away—proceeded to land gear up, doing extensive damage to an otherwise unblemished aircraft. During the inquiry he was asked how such a thing could happen after his successful effort to make it back to the airport in the crippled aircraft. He told the board of inquiry that he was under enormous stress and couldn't think clearly because of a loud horn blasting in the cockpit as he approached the runway. It was the gear-up warning horn.

Interestingly, landing gear was an afterthought. The Wright Flyer used skids and a rail system. Dr. Samuel Langley's ill-fated Aerodrome was to be launched from a houseboat floating down the Potomac River, then landed on the water. Putting tires on an airplane wasn't seriously considered until 1909, when the U.S. Army purchased its first aircraft and specified that it should have landing gear with inflatable tires.

Modern landing gear, while certainly more sophisticated than ever, is still fairly straightforward. Nonretractable landing gear are still common on many light, general-aviation trainer aircraft, but for the most part aircraft have hydraulically operated landing gear such as the Piper Seminole hydraulic gear system shown in Figure 10-1. There are still one or two pneumatically driven systems, and a few light aircraft use electric motors.

Functionally, most retractable landing gear systems work the same way—even an amphibious landing gear system such as that used on the Cessna Turbo Stationair 6 shown in Figure 10-2. They generally have a two-position switch for GEAR UP and GEAR DOWN. Typically there is a single red light to indicate that the gear are not down-and-locked.

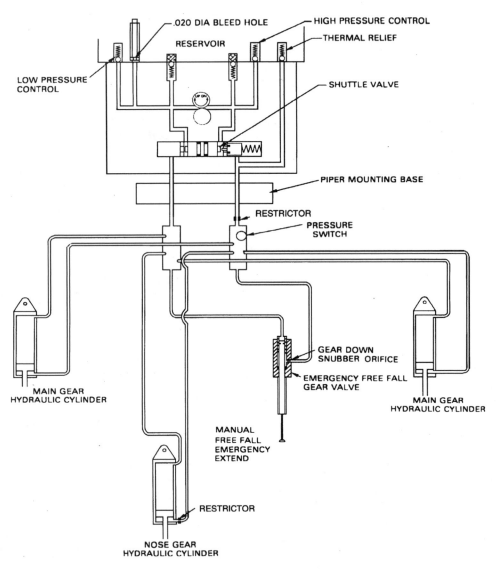

Fig. 10-1. *PA-44-180 Seminole landing gear hydraulic system.* (Courtesy of The New Piper Aircraft Company)

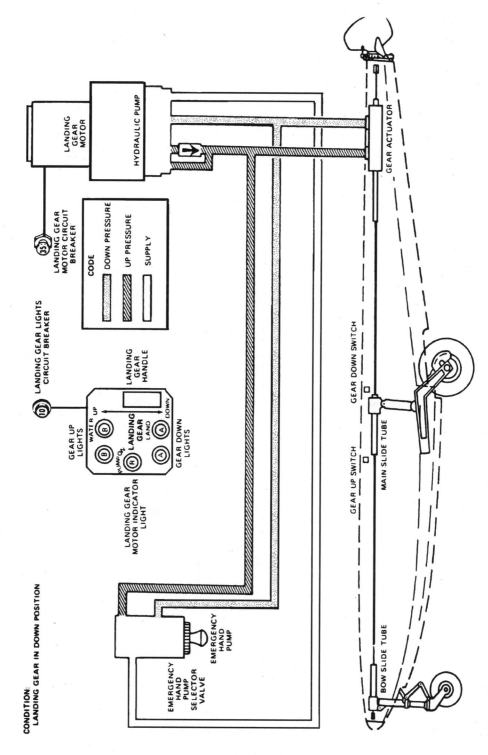

Fig. 10-2. *Cessna TU206G Turbo Stationair amphibious landing gear system.* (Courtesy of Cessna Aircraft Company)

CONDITION:
LANDING GEAR IN DOWN POSITION

LANDING GEAR MOTOR

HYDRAULIC PUMP

35

LANDING GEAR MOTOR CIRCUIT BREAKER

CODE

DOWN PRESSURE

UP PRESSURE

SUPPLY

10

LANDING GEAR LIGHTS CIRCUIT BREAKER

LANDING GEAR HANDLE

WATER UP

UP

B

GEAR UP LIGHTS

DOWN

LAND

A

LANDING GEAR

R

PUMP ON

B

A

GEAR DOWN LIGHTS

LANDING GEAR MOTOR INDICATOR LIGHT

EMERGENCY HAND PUMP SELECTOR VALVE

EMERGENCY HAND PUMP

GEAR ACTUATOR

GEAR DOWN SWITCH

GEAR UP SWITCH

MAIN SLIDE TUBE

BOW SLIDE TUBE

221

Normally, this indicates that the gears are retracted if the gear switch is in the GEAR UP position. However, if the gear switch is in the GEAR DOWN position and you continue to have a red light, it is often difficult to know exactly what that means.

It could mean nothing at all has happened with the gear since you put the switch down, indicating a probable total system failure. It could also mean that the gear began to extend but one or more are not in the fully extended down-and-locked position. A major clue to the status of the gear is whether or not you had an aerodynamic change when you put the switch down. If not, it's likely none of the gear have moved. If there was an aerodynamic effect, then you have one or more landing gear that have failed to deploy and lock. That's a tough situation and a tough call. If possible, fly by a tower and have them take a look at your gear. And be very sure to follow the emergency procedure outlined in your aircraft POH.

The other side of the coin is forgetting to extend the gear on approach to landing. Landing gear systems incorporate lights and/or audible alarms that activate when you put the aircraft into landing configuration but fail to extend the gear. Generally the criteria for the alert to activate includes such things as reducing power below a specified manifold pressure and extending flaps into the landing setting. Your POH will fill you in on the details, but the important point to be made is that you should always use a checklist for every stage of a flight.

Forget the embarrassment of landing gear up; that's trivial. Consider the consequences. You will almost certainly damage the retracted landing gear; you will definitely ruin the belly paint and almost certainly cause structural damage to some part of the belly. You'll knock off all the belly-mounted antennas, and—worst of all—you'll stop the props, causing prop and crankshaft structural damage. Use your checklist!

TIRES

Kick the tires and light the fires is a half-joking expression that has carried over from the early days of flying. It does point out, however, that tires are an important preflight item, even to the devil-may-care aviator of old. Though dormant from takeoff to landing, aircraft tires can give ground operation a whole new meaning if they don't do their job correctly. They are designed to give a comfortable ride to both the airframe and passengers, to provide easy ground maneuvering, and to maximize braking.

Pilots usually don't think of tires as important control surfaces, but imagine a blowout on takeoff at 75 knots. The immediate and overwhelming loss of control could be devastating, yet tires probably are the most neglected part of the airframe preflight.

Anatomy of a Tire

There is a substantial difference between automotive and aircraft tires. As unlikely as it may seem, they are actually designed to meet almost opposite requirements. Automotive tires are required to run down a highway at high speeds for a long time, carrying a fairly heavy but constant load. As such, automotive tires require very little flex, called *deflection*, so the manufacturers design them for a continuous deflection of approximately 12–14%.

Aircraft tires, on the other hand, actually spend very little time rotating, but when they do it can be at speeds far in excess of those that most cars will experience. Plus, the tire is subjected to going from a no-load situation to a very heavy load instantly, often with substantial impact, while spooling up from zero to often over 100 mph. Manufacturers design aircraft tires to have a deflection capability ranging from about 32 to 35%.

Main Tire Components

There are four main components to an aircraft tire, which may be seen in Figure 10-3. They are the bead, carcass (cord body), tread, and sidewall.

In the world of tires, the bead is king. It is the bead that ultimately bears the brunt of all the forces applied to the tire. Plus, the bead anchors the carcass and provides a method to mount the tire on a wheel.

A tire's carcass is composed of plies (layers) of rubber-coated nylon cord fabric. Cut on a bias into strips, the fabric's cords run at an angle roughly 45 degrees to the length of the strip, which extends across the tire and around the bead. As each ply is put on the tire, it is oriented so that its cords cross the subsequent cords at about 90 degrees, giving added strength and balance to the carcass.

The tread is the surface on the outer circumference of the tire that principally contacts the ground. Various patterns of grooves make up the tread, whose purpose is to increase

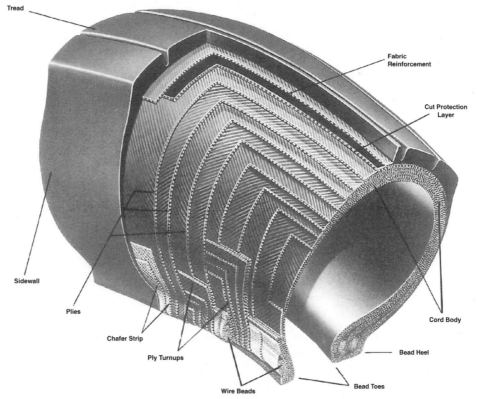

Fig. 10-3. *Aircraft tire cross section.*

the tires traction on the ground. There are a number of different types of tread, such as all weather and ribbed. The all-weather tread is a diamond-shaped pattern grooved into the surface of the tire and is usually found on aircraft that operate often on grass or dirt strips.

Ribbed tread tires are by far the most common today on light, general-aviation aircraft. They are characterized by parallel, continuous grooves that run along the outer circumference of the tire. Designed specifically to operate off hard-surfaced runways, water will go up into the grooves, allowing the tire surface to touch the ground for better traction. Ribbed tires are known not only for their excellent traction but also for their long service and excellent directional stability.

The sidewall is actually a protective rubber covering that goes from the tread to the bead. Its purpose is to protect the carcass exposure to moisture, scrapes, scuffs, cuts, or bruises.

Tire Wear

Tires are designed and built to be remarkably strong and flexible, so catastrophic failure is not a common problem, though occasionally it does occur. Excessive wear, on the other hand, is the biggest problem with tires.

Ask the average pilot what is the most significant factor reducing tire life and you'll probably get the answer, "Hard landings." There is no question that hard landings make an impression on pilots, passengers, landing gear, and airframes, but hard landings are less of a problem for tires than pilots may think. Aircraft tires, much more flexible than their automobile counterparts, are designed to flex as much as 35% from their original shape. This allows for landings that are a lot harder than the average pilot can handle!

The real killer for tires is heat. The more flexible the tire, the greater the internal heat generated as a result of stress and friction. If you have the slightest doubt about that, drive your car a half mile down your street at 60+ mph, and then, while the nice officer is writing the ticket, feel the tires. The heavier the vehicle, the more heat that will be generated; the more heat, the shorter the life of the tire.

There are several ways to minimize the heat problem, but they tend to be unpopular with pilots:

1. Always keep tires properly inflated.

2. Always taxi as slow as practical. The slower you go, the less stress on the tire and the less it will deflect, which translates into less heat buildup.

3. Keep ground maneuvering to a minimum. When possible, plan to land so you will roll out close to your destination on the airport.

4. Reduce, and if possible eliminate, braking. Not only does this save tires, it also saves brakes. Braking causes friction between the tire and the ground, which leads to a fast buildup of heat. Also, the brake itself creates a significant amount of heat, part of which will transfer to the tire. A touchdown and long rollout without brakes is preferable to a short ground run with brakes, assuming there are no obstructions looming at the end of a short runway.

There is yet another reason to avoid hard braking: tread wear. Friction means something has to give, and tires are much softer than runways. Think of what happens when you're grating a potato to make hash browns for breakfast and your thumb gets in the way: presto—meat and potatoes. In short, the rougher surface will always win. Every time you apply brakes in an airplane or car, you scuff off some of the rubber tread. Tires with little or no tread must be replaced. Figure 10-4 shows an area toward the middle of the tire where the tread has been worn away. Also, hard braking leads to skidding.

The reason the wheels don't lock up immediately when you apply the brakes is the momentum of the airplane combined with ground friction on the tires greater than the force the brake applies to the wheel. If you touch down on a runway with areas of low friction, such as wet spots or patches of ice, and attempt hard braking, one wheel eventually will cross an area of lower friction than the other will. When that happens, the wheel encountering low friction will lock up while the other will continue to turn. Aside from a slight lack of control, this does not produce a problem. It is when the locked-up wheel leaves the patch of ice that the trouble begins. At that instant, the brake-locked, zero-RPM wheel is being dragged over a high-friction surface by the momentum of the airplane.

Speed is a nemesis of tires; whenever safely possible, operate as slowly as you can. Excessive speed produces more internal heat and requires more braking. Touch down as slowly as practical for the conditions; do full-stall landings, then taxi slowly to the tiedowns. While taxiing, be particularly aware of turning. Tight turns are a significant cause of tread wear, especially pivoting about one wheel using the brakes. This creates a terrific

Fig. 10-4. *Center tread worn to bald condition.* (Photo by author, courtesy of Frasca Air Services)

sheer force, not only on the main gear tires, but also on the nosewheel. Tight turns force the nosewheel to flex, scuffing the sidewall area and causing the tire to go out of balance. This is one of the primary causes of nosewheel shimmy and greatly reduces tire life.

Tight turns also put unnecessary stress on tire casing, beads, and sidewalls. They produce flat spots, which put the tire out of balance and cause it to thump and the airplane to bounce while taxiing. Tight turns on gravel can cause a piece of stone to screw into the tire and puncture it. Often, because of the nature of the puncture, the flat tire isn't discovered until the next trip to the airport, such as the aircraft shown in Figure 10-5. If a tight turn is impossible to avoid, allow the tire that is inside the turn to make as large an arc as possible to minimize potential damage.

Inflation Pressure

Always keep tires inflated to the value specified in the pilot's operating handbook (POH) rather than the tire manufacturer's specifications. While this may sound contrary to the

Fig. 10-5. *Pivoting turns on gravel can lead to flat tires.*

usual practice in aviation, the POH accurately reflects the actual tire loading for that specific airplane. The specifications from the manufacturer do not take into consideration the specific tire loading value. In short, the airframe manufacturer knows the specific application; the tire manufacturer doesn't. The single most important preventive maintenance the owner can do is to keep the tire pressure correct at all times. This will produce more landings per tire than anything else will.

When new tires are installed, they typically are inflated without the aircraft weight acting upon them. Because aircraft weight deflects the tire, approximately a 4 percent increase in inflation pressure is required. New tires should be allowed to sit for 24 hours after installation, so they can stretch and adjust to the rims. It is also important to check tire pressure daily during the first week because of probable air leakage.

Tire pressure should be checked with an accurate dial gauge prior to every flight, a preflight item that is seldom done by the average pilot. The heavier the airplane, the more critical this check becomes. It must be done on a cold tire, at least four to five hours since the last use, because a hot tire gives a high reading. When going on a long cross-country flight, tires should be inflated for the coldest condition to be encountered. Ground temperature changes of 50 degrees F or more require greater inflation to compensate for the lower temperature. A good rule of thumb is every 5 degrees Fahrenheit temperature change yields a 1% change in tire pressure.

Improper inflation causes uneven tread wear. An underinflated tire will be subject to excessive wear on the shoulder. This is the worst possible condition because it scars the sidewalls and shoulder as it rubs against the tire-rim flange. The result is faster heat buildup. In extreme cases, tube tires may slip around the rim, shearing off the valve stem. Overinflation, on the other hand, causes excessive wear in the center of the tire, reduces traction and ground-handling ability, increases landing distance, and makes tire treads more vulnerable to cuts and nicks.

Most cuts and nicks can be avoided if the pilot will use caution and always watch the taxiway ahead. Foreign objects on runways and taxiways are a prime cause of cuts and nicks and should be reported immediately. If they are unavoidable, either get out of the airplane and remove them yourself, or call unicom/ground control and ask for assistance. With the price of tires today, patience has never been a greater virtue.

Other common problems are potholes, large surface cracks, and dropoffs between tiedown areas and taxiways. These should be avoided when possible because they really can take a chunk out of a tire. The only solution, if they are unavoidable, is to take them as slowly as possible. Be particularly careful of dropoffs. Not only are dropoffs a problem for the tire, but if deep enough, you may also catch the prop.

The part of the tire that makes contact with the surface is called the "footprint." A tire can be compared to a running shoe: both have tread gripping the ground. The less tread that touches the ground, the less the grip. The groove between treads allows water to pass under the tire without losing its grip on the surface. Skidding, a situation where one tire locks up while the other continues to turn, can result in serious problems. Skidding leads to blowouts, loss of control, and significantly increased landing rolls.

Hydroplaning

There are three types of hydroplaning: dynamic, viscous, and reverted rubber.

Dynamic hydroplaning requires 1/10th inch or more of standing water on the runway. Excessive tire tread wear, insufficient groove depth, and overinflation exacerbate the problem as the tire alternately rolls on the dry surface and skis on the wet. Pilots who have taken off or landed during heavy rain have probably experienced this phenomena, though they may not have realized they were hydroplaning. In the airplane, it feels as if the airplane is alternately sliding and jerking, typically from side to side.

Viscous hydroplaning is probably the most common of the three types. It occurs on a smooth runway or one on which there are rubber deposits present, such as in the touchdown area. It requires only a thin film of water that the tires are unable to penetrate, causing a partial loss of contact with the runway. It can happen at a speed significantly below that which you normally would anticipate hydroplaning to occur—even at a fast taxi!

Reverted rubber hydroplaning is a bit more complicated. The necessary conditions are a wet runway surface and a skid in progress. When the tire locks up because of hard braking on a slick surface, the resultant friction generates heat and the tire begins to smoke. Rubber debris collects under the tire, causing water to build up in front of, and underneath, the tire. The heat turns the water into steam and the tire actually rises up and floats on the steam. There is no runway contact, a severe loss of control, and, yes, it really can happen in light aircraft as well as heavies.

There is a simple formula to determine at what speed an airplane will hydroplane. It is 8.6 times the square root of the tire pressure measured in pounds per square inch. That will give you the lowest entry speed; however, once hydroplaning has begun it can continue at lower speeds! For instance, the Cessna 172Q nosewheel holds 45 psi, and the main gears hold 38 psi. That means that the nosewheel will begin hydroplaning (if the conditions exist) at 50 knots and the main gear at 46 knots.

A Beech 58A Baron nosewheel has 55 psi; the main gear has 52 psi. Hydroplaning is at 56 and 54 knots, respectively. Obviously, the best plan of attack is to try to touch down below those speeds if there is standing water or other conditions conducive to hydroplaning. Further, it is advisable to not use brakes until the aircraft has decelerated below the calculated hydroplaning speed. Instead, use aerodynamic braking until reaching that speed.

Remember that hydroplaning can happen anytime you are at or above the speed, whether you are landing or taking off. Because the grooves between tire treads play an important role in getting water out from under the tire, it is important that they be checked on each preflight. The best method is to use a manufacturer approved depth gauge, but at least check them visually.

Preflight

There is not a lot of preflight associated with tires. Probably the most important thing is to look at their overall appearance and condition. Look carefully for cuts and nicks. Check cold tire inflation against the POH's recommendation for tire pressure. Finally, tread should not be worn down to the point where patches of the tread disappear into a smooth area. If they do, it is time to change the tire.

Preventive Maintenance

Keep tires clean. Often overlooked is the effect of contaminants on tires. It is not uncommon to see an airplane tied down with a tire sitting in a pool of oil or hydraulic fluid. It is important to understand that all mineral-based fluids deteriorate rubber. Gasoline, oil, hydraulic fluid, grease, and tar are pure poison for tires. Any foreign substance on a tire should be cleaned off with soap and water. If that doesn't do the trick, take it off with gasoline, then remove the gasoline with soap and water.

Protect tires from the elements. Tires on airplanes that normally are tied down outside have an additional problem: weather checking. Harsh outdoor elements eventually lead to checking of tires, little hairline cracks. The tires essentially dry out and become somewhat brittle. Provided the plies are not visible, this should not cause any significant problems. Similarly, sunlight and static electricity in the air will convert oxygen to ozone, which also attacks rubber and causes aging. This can be avoided by putting flexible, aluminum-coated covers over your tires when the airplane is not in use. Incidentally, if you store your airplane in a hangar, don't breathe a sigh of relief too quickly. If there is an electric motor running inside the hangar, it also will interact with oxygen, create ozone, and go to work on your tires.

Replace tires when they are worn. Most tire manufacturers recommend replacing tires whenever the tread is worn to the base of any groove or to the minimum depth specified by the airframe manufacturer. If hydroplaning seems to be a persistent problem, tires with only 2/32 to 3/32 inch of groove depth should be removed. But remember, this is only a guide; there are many variables to tire replacement schedules.

Always remove a tire if there is any ply fabric exposed anywhere—ALWAYS! Also, if a cut or crack is greater than 50% of a tread rib, if it goes to the base of a tread groove, or if it undercuts a tread rib—remove it. Finally, bulges in any part of the tread, sidewall or bead area, indicate separation and permanent damage to the tire and require removal.

TYPES OF BRAKES

Stopping the airplane is what brakes are all about; friction is how it's done. Orville and Wilbur had the simplest friction brakes, a tailskid that dragged along the ground. Such an arrangement was more than adequate in those days because airplanes had very slow landing speeds and the airports were nothing more than a convenient field. Of course, every time an airplane landed, it tore up the ground somewhat, but with so few airplanes, the low cost of crops, and a society that had yet to discover the concept of suing everyone for anything, that really didn't matter.

Ground maneuvering was a bit of a trick; every time you wanted to turn, you had to rev up the engine so the propeller would send a blast of air back over the tail, lifting it up. Once the weight was off the skid, the rudder was used to turn the airplane as necessary. Nothing fancy, just basic applied aerodynamics.

Chapter Ten

It wasn't long before flying became popular and pilots started thinking of practical uses for the airplane. First, they thought of carrying one passenger, then two, then a package or two, mail, and as a result airplanes got bigger and heavier. Approach speeds increased, landing distances grew longer, ground maneuvering became more difficult, and hard surface runways began to appear. The need for pilot-controlled brakes became obvious. It was a logical step to borrow the idea from the automobile brake and adapt it to the airplane. And that's just what was done; the standard drum-and-shoe type brake began to appear on airplanes.

Aircraft ground handling and control improved, but airplanes continued to get heavier and faster. Brake fade, the drawback of the automotive brake, began to rear its ugly head in aviation. The friction to stop the airplane was caused by an asbestos-lined metal shoe that wedged against a rotating cast iron drum attached to the inside of the wheel. In principle, it worked well; in practice, it was another thing completely.

As engine power, gross weights, and speeds increased, the demand on the brake increased. In fact, the modern brake has several jobs. It provides sufficient friction to stop the airplane on landing, holds the airplane during engine runup, absorbs the kinetic energy of the wheel and converts it into heat, and sheds the resultant heat as efficiently and quickly as possible. The old shoe and drum just couldn't handle that much work. The heat became high enough to cause the rotating cast-iron drum to expand away from the shoe, which resulted in a dramatic reduction of brake effectiveness. A better method had to be devised, and after some serious effort the disc brake was designed.

Probably one of the simplest systems in most light aircraft, the brakes are usually activated by a dedicated hydraulic system. The major components, shown in Figure 10-6, are rudder pedal/toe brakes; master cylinder; the hydraulic tubing brake line; disc; and brake housing, which includes the piston and linings, also known as pads. The toe brakes are integral to the rudder pedals, though a few aircraft use either a hand brake or independent heel brakes. When using toe brakes, it is important not to confuse rudder pedal deflection with pushing on the top half of the pedal for brake application. The two functions are independent of each other, as Figure 10-7 illustrates. The left toe brake operates the left main gear brake while the right toe brake operates the right main gear brake, making differential braking possible. There is one master cylinder for each toe brake, and its purpose is to translate foot pressure into hydraulic fluid pressure through the brake line to operate the wheel brake.

Single Disc Systems

Virtually all modern, light aircraft use a single disc brake system. The system is composed of two large friction-producing calipers, one on each wheel, which are controlled by the brake pedals. A caliper works by clamping down on a disc attached to the rotating wheel.

One reason for the popularity of the single-disc brake system in light aircraft is its simplicity, as seen in Figure 10-8. It consists of a brake unit housing, typically made of aluminum or magnesium alloy, attached to the landing-gear strut. Within that housing is a steel disc that is rigidly fixed to, and rotates with, the aircraft wheel. The purpose of the

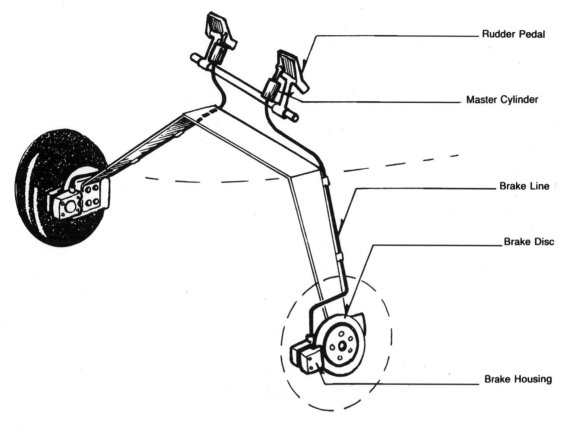

Fig. 10-6. *Major components of aircraft brakes.*

disc is to provide a gripping surface for the two brake linings, which are mounted in the brake unit housing. The brake linings are mounted in such a manner that they will evenly apply pressure to both sides of the disc.

When your foot presses on a pedal, let's say the right one, the pressure is transmitted by the hydraulic fluid to the piston in the right brake housing. The piston responds to the increased pressure by pushing against a lining, which moves horizontally toward the disk. The lining causes friction by pushing against the brake disc, which in turn pushes against a stationary lining on the other side, causing yet more friction. Figure 10-9 shows the disk sandwiched between the linings. Not only does this provide twice the stopping power, but because of equal friction on both sides of the disc, it also causes even disc and lining wear. The whole process is nothing more than a sophisticated version of the bicycle hand brake.

Disc brakes are usually made of forged steel. For most airplanes (those that are flown regularly and not excessively exposed to a corrosive atmosphere), they are trouble free. To paraphrase a famous saying, however, "Steel will be steel," and there are some potentially significant problems.

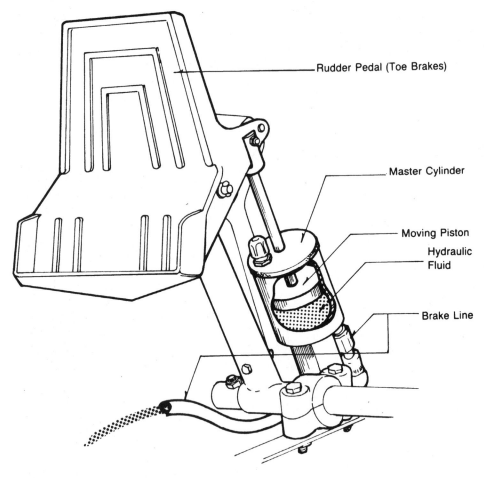

Rudder Pedal (Toe Brakes)

Master Cylinder

Moving Piston

Hydraulic Fluid

Brake Line

Fig. 10-7. *Standard foot brake.*

If your pride-and-joy flies less than 200 hours per year and especially if it is exposed to unusual amounts of moisture, salt, or industrial chemicals, you may have a cancer slowly eating away at its discs. Such an airplane sits idle for more than 8500 hours per year, which means not enough time is spent rubbing corrosion and rust off those discs. Not to worry—if you're in that category the answer is chromed discs.

Chromed Discs

When most of us think of chrome we think of the flashy trim on our cars, but chrome discs aren't flashy. Far from being a show item, chrome discs are rather dull in appearance; however, they are designed to prevent corrosion and rusting. For a number of years there were a few mechanics who would take your old, rusted, pitted brake discs, completely resurface them, and finish the job with a chrome treatment. Pilots who used them began swearing by chromed brakes. However, there were many skeptics. Most of them imagined that the chrome was too slick and wouldn't produce the same braking friction as the traditional discs. Well, they were wrong, as tests have proven that the coefficient of friction for chromed discs is comparable to regular discs. Today, discs are no longer refinished in chrome. Old discs are just that—old discs. Chrome discs are made new, and while it's true that a set of chrome disc brakes is more expensive, the cost is more than offset by the advantages. Chromed discs have significantly longer life, reduced overall maintenance costs, and shorter downtime. Most single-engine aircraft now offer the option of chrome discs as original equipment, and virtually all singles can have them installed as service replacements.

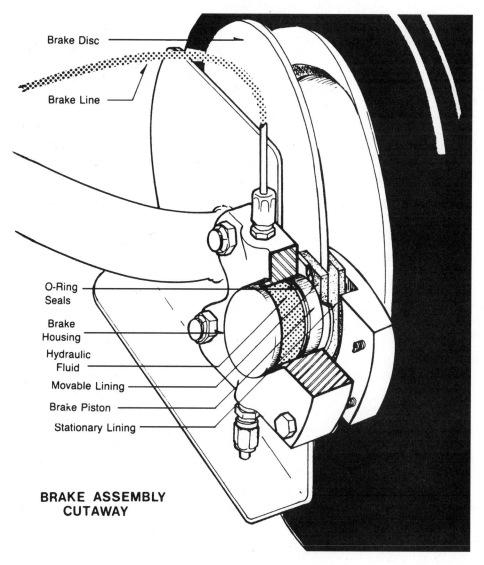

Fig. 10-8. *Single disc brake system.*

Fig. 10-9. *Brake pads sandwich disk in between.* (Photo by author, courtesy of Frasca Air Services)

Brake Fluid

The prime mover of the brake system is its hydraulic fluid. Without it, there's no stopping you. While it is fairly obvious that brake fluid is responsible for transmitting pressure and energy, it does serve other functions, too. It lubricates the moving portions of the system it comes in contact with, and it aids in cooling the working parts.

While there are several types of hydraulic fluid, most light aircraft use a mineral-based fluid, which consists of a high-quality petroleum oil. A common type is MIL-H-5606, which is less corrosive than some of the others and can be identified by its reddish color. However, due to its petroleum base it is flammable and caution should be exercised in using and storing it. It is very important that you check your owner's manual to find out exactly which type your aircraft uses, since mixing different fluids may render the system useless. Some hydraulic fluids can actually eat the rubber seals of incompatible systems! Therefore, it is a good idea to mark the system's filler cap with the type of fluid to be used so no one will accidentally add a different kind. When storing fluid, be particularly careful to protect it from possible contamination by dirt. Particles of dirt can render a system inoperative almost immediately. Try to pick a time to replenish the system when there isn't a dust storm looming up at the edge of the airport; dust and dirt in the fluid are a significant cause of hydraulic system failure!

Even before you start the engine you can learn whether or not you have sufficient fluid in the system. If you step on the brakes and there is pedal movement, you have fluid.

If, on the other hand, the pedal lays all the way back and there is no movement, you are out of fluid. Take a look at the floor to see if there is hydraulic fluid on it, an indication of a leaking master cylinder. If there isn't, step outside and check the ground around the tire. You'll be looking for an indication of a leak in the hydraulic line, the brake piston "O" ring, or the hydraulic line fitting where it enters the brake housing. Leaks of this nature automatically call for the expertise of a mechanic. Another potential problem easily observed on preflight would be a twisted hydraulic fitting line to the brake housing. If the line has a kink, it will impose a side load on the brake housing, and uneven lining wear will result in potentially reduced braking power and a shorter life for the linings.

Probably the most dreaded brake problem one can imagine would be to press down on the pedal during the landing roll and get little or no response. Assuming you actually have pedal movement, indicating there is fluid in the system, the probable cause is dirt. This situation is avoidable if you keep your brakes clean. The culprit is likely to be a buildup of dirt on the through bolts (also called guide pins) that allow the movable lining to slide back and forth in response to the piston. Incidentally, this same problem can also cause the brakes to drag if the unit should freeze up when the linings are in contact with the disc. You'd recognize this by a scraping noise when you taxi. To check this during preflight, grasp the brake housing with your hand and try to twist it. If it is free, it should move slightly; if you can't get it to budge, it is probably frozen in place because of packed dirt.

Disassembling a unit for cleaning is very simple, but the first time it is always best to get a little dual instruction from someone who has experience. After you clean the bolts you can help prevent the problem from occurring again by lubricating them, but be very cautious what you use for lubricant. Oil, a tempting choice, will only make the problem worse by collecting dirt and holding it in place. Instead, you should use graphite, Dri-Slide, or silicone spray, any of which will lubricate without acting as a sticky surface to attract and hold dirt.

Getting into the airplane and stepping on the brakes during preflight may tell you if you have fluid, but it won't give you an indication of the status of the brake linings. The brake pedal will feel fine right up until the second the disc wears off the rivet heads and the linings fall out of the brakes. There is only one way to assure that the linings are good; get down on your knees and visually inspect them. New linings are approximately 0.25 inch thick and should be replaced when they are worn to a thickness of 0.10 inch or less. While you are down there looking at the linings, check for the presence of grease or oil on the linings or disc surfaces. A light coating of either will reduce braking effectiveness greatly.

Troubleshooting

The chief flight instructor of a major university flight program related to me a problem that surfaced when he started getting a lot of bills for brake relining. Their fleet of aircraft was getting new brakes far too often and the culprit was simple negligence; students were riding the brakes.

Chapter Ten

There is a real potential for wearing out the brakes when the rudder pedal serves the dual purpose of rudder control and braking. Unfortunately the only cure is never to taxi with your feet flat on the pedal; instead, put your feet on the floor and toes on the bottom part of the pedal. It's a simple matter to slide your foot up and press the brake when you need it. If you feel you must keep your feet on the brakes while taxiing to be able to instantly react, you are probably taxiing too fast. Riding the brakes causes unnecessary, and rapid, wear of the linings. During taxi, you should seldom, if ever, touch the brakes. Using them to keep your taxi speed slow is an indication of carrying too much power. The automotive equivalent is to pull out of your driveway by flooring the accelerator with one foot and standing on the brakes with the other.

Another common problem is using the brakes to turn. Generally it is a sign of poor planning by not mentally staying ahead of the airplane. Rather than using brakes for turning in the multiengine airplane, there is the advantage of differential power. If you desire to turn left, you just idle power on the left engine and slightly increase power on the right while using left rudder.

Operational Considerations

Never raise flaps on the landing roll! That simple rule, ingrained in countless students over the years, has absolutely nothing to do with stopping the airplane. Rather, it stems from the sad fact that many pilots have inadvertently retracted the gear instead of the flaps during landing. Never mind the landing gear squat switch; it fails. Never mind that the gear switch is round and the flap handle is flat; preoccupation tends to cover that up, and never mind that they are typically located in totally different places. An alarming number of pilots have overcome all of these obstacles and managed to retract the gear on the runway. Therefore, most instructors tell you to never raise flaps on the landing roll! Unfortunately, that is contrary to maximum-performance landings.

Okay, let's be honest. If you are landing your Cessna 152 at Chicago's O'Hare International Airport and you have two miles of runway, who cares? Well, assuming there isn't a DC 10 right behind you, nobody. In fact, it is prudent to use little or no brakes on landing; brakes wear out when they are used. Given the opportunity, I will roll out to the end of the runway to save on brakes. What happens when you are forced to make an actual short-field landing? In that situation, everything you have learned about landings, and routinely practiced, will give you the most inefficient technique. For maximum performance landings, you should touch down slightly nose-high, at the slowest safe speed for your gross weight, with full flaps. This configuration will be effective to approximately 60% of your touchdown speed. For example, if your touchdown speed is 100 knots, aerodynamic braking will be more effective than friction down to approximately 60 knots. The reason: The wings will still be creating sufficient lift to reduce weight on the tires, which minimizes friction braking. Once you reach 60% of your touchdown speed, let the nosewheel contact the ground, retract the flaps, and begin to apply smooth, maximum brake pressure without allowing the tires to skid.

Installing New Brakes

As the saying goes, all good things must come to an end. In the case of brakes, the end applies to linings. When your mechanic has finished the job, the new brake linings must be conditioned before use. All too often pilots, and even mechanics, are unaware of the need for breaking them in. There are actually two conditioning procedures. The one you use will depend on the type of linings used on your airplane.

Asbestos-based organic composition brake linings must have the resins properly cured before the brakes are actually used. Failure to properly condition brakes may result in carburizing the linings with a single hard application, preventing a good braking coefficient and significantly shortening the life of the linings. The procedure is very simple. Taxi the airplane at a speed of 25–40 mph; then, using a light braking effort, gently bring the aircraft to a full stop. Wait at least two minutes for the brakes to cool down. Then repeat. This procedure should be done a total of six times, each time allowing at least two minutes for the brakes to cool. Keeping the taxi speed between 25 and 40 m.p.h. and using light braking generates sufficient heat to cure the resins, but not enough to carburize them. When strictly adhered to, this procedure virtually guarantees properly cured brake linings that should get about 100 hours of taxi time.

The iron-based metallic composition brake linings require a glazing process after installation. Here a simple procedure is used, with significantly different numbers from the organic method. Taxi speed should be 30–35 knots (excess speed may cause overheating and disc warping) with a hard, full-stop braking application. Then immediately go back to the 30–35-knot taxi speed and repeat the procedure once more, for a total of two times. Unlike the organic break-in procedure, you do not want the brakes to cool down between the two taxi runs. If the procedure is done correctly, the high spots will wear off the linings and the result will be a flat, smooth surface. It is a good idea to check the linings during preflight, and if they begin to appear rough or grooved, repeat the conditioning procedure. One potential problem requiring reconditioning is wearing the glaze off the linings. This is the result of frequent, light brake applications during taxi, or worse—riding the brakes.

Preventive Maintenance

Hydraulic fluid poses the greatest potential problem. But without it, there are no brakes at all. It is important to maintain the integrity of the lines by making sure they are not kinked or chafing against another part. Routinely check the hydraulic fluid level, and on every preflight look for puddles of fluid on the ground or streaks of reddish fluid on the airplane. Hydraulic leaks should be referred to a mechanic.

Dirt is the nemesis of brake systems. If allowed to get into hydraulic fluid, even grit can cause destruction of seals, erosion of moving parts, and total brake system failure. Allowed to collect around the brake housing and anchor bolts, dirt may cause the brake to "freeze up." Keep the bolts clean and lubricated with Dri-Slide, graphite, or silicone spray—never oil! One word of caution: do not loosen or attempt to remove the anchor bolts. The procedure for cleaning them is similar to that required to reline the brakes, and while not difficult, it should not be undertaken without some instruction.

Oil or grease on the surfaces of brake discs or linings will cause a significant loss of braking friction; if either is present, remove it with solvent. While you are checking the linings, make sure the surfaces are evenly worn and they are greater than 0.10 inch in thickness. Anything less requires replacement.

With brake system problems, like anything else, you want to give the mechanic as much information as possible. Dragging brakes are indicated during taxi by a squeaking or scraping sound from the brake and if serious enough may show a tendency for the airplane to pull to one side. When describing problems of insufficient braking, things to note would include whether or not the brake pedal feels "spongy" (indicating possible air in the hydraulic fluid) or if the pedal is flat (indicating no fluid). If you apply even pressure on the brakes and the airplane tends to pull toward one side, you have differential braking, which indicates that only one of the two brake systems is malfunctioning.

If there is little or no response to the pedal, the mechanic will want to know what the surface conditions were when you experienced the problem. I recall some years ago giving a Grumman Tiger checkout to a CFI who had never flown one. It was a typical Illinois early winter night with a wet ramp and low freezing level. We taxied out, took off, and conducted an uneventful flight. I remember thinking to myself as we entered the pattern, there was a slight crosswind and the pilot had no experience in a caster gear airplane (the Tiger does not have a steerable nosewheel). Because I knew he was a very competent pilot and instructor, I decided to remain relatively docile during the early stage of the landing to allow him plenty of time for self-discovery (not to mention recovery). Immediately upon touchdown the airplane began to veer toward the right, but I remained cool because it was the educationally sound thing to do. I could feel the learning taking place; so much for educational theory. About the time he yelled, "You got it," we hit it—the runway light, that is.

I learned something very important that night. When you taxi through water on the ground, then fly above the freezing level, the wet brakes can freeze up solid. The cost of that little educational experience was one runway light, a tire so squared off it would stand upright in a hurricane, and one chief flight instructor with egg all over his face. The end result was rather drastic, but surface conditions can also contribute to brake problems, so they are worth noting.

Low fluid indicates a leak, but not where the leak is located. Puddles of hydraulic fluid discovered during a preflight should be carefully noted with respect to where they are relative to the aircraft. This can lead the mechanic directly to the source of the leak—something that may not be easily found otherwise.

Finally, you should keep track of the condition of the discs. Any of the following irregularities of the disc faces should be reported to the mechanic: warpage, irregular wear, scouring, grooves, pitting, rusting, or corrosion. The brakes—perhaps more than any other system—are directly related to the treatment they receive by the pilot. If used correctly, preflighted routinely, and cared for properly, they will give years of useful, surprise-free life.

11
Environmental Systems

LARGE TURBINE-ENGINE AIRCRAFT, WITH THEIR SOPHISTICATED environmental systems, are able to isolate occupants from outside weather. Unfortunately, the light airplane traveler does not fare quite so well.

For one thing, the airlines literally have a trick up their sleeve: It's called an airport jetway. Even the most sophisticated air conditioning and heating systems can't keep a cabin comfortable if it has a door open to the world. The jetway cleverly forms a sleeve that connects the airplane to the terminal, allowing maintenance of cabin temperature.

At terminals without jetways, a quick turnaround reduces the time cabin doors are open. Then, too, airline cabins are long and have several bulkhead, minimizing the flow of air out the door. The relatively small size of the average general aviation fuselage makes the inside temperature sensitive to open cockpit doors. And the inadequacy of heating and cooling systems precludes the luxury of a stable cabin environment unless you are able to load and unload passengers in a temperature-controlled hangar.

TYPES OF AIRCRAFT HEATERS

Aircraft powered by turbine engines have a ready source of heat, the turbine engine itself. Reciprocating engine-powered aircraft do not have that available so cabin heating must come from another source. There are two types of heaters for light, general-aviation airplanes: an exhaust-manifold heater and a combustion heater.

Exhaust Manifold Heater

The exhaust manifold heater, used exclusively be reciprocating, single-engine aircraft, is the simplest. Working on a simple heat-transfer principle, a shroud is placed around the engine exhaust stack as illustrated in Figure 11-1. Fresh outside air is forced by ram pressure through the shroud and around the exhaust stack. The stack isolates the exhaust gas from the fresh air but allows the heat to transfer. To vary the temperature the pilot controls a source of additional outside air that mixes with the heated air to cool it to the desired temperature.

The main advantages of the exhaust manifold heater are simplicity, low maintenance, and virtually no reduction in flight performance. It doesn't consume fuel, reduce engine power, or decrease airspeed by any detectable amount. The disadvantages, however, are obvious to anyone who has ever flown in a lightplane during winter. An exhaust manifold heater is a very ineffective system on the ground because minimal ram air moves through the shroud. Not only does this mean the cabin tends to stay cold during ground operations, it also means poor windshield defrosting. Because windshield defrosting is accomplished by the rechanneling of cabin heat, it is common for pilots to taxi and take off with fogged windshields, peering through small circles smeared away with the back of their hands, leaving nose prints on the acrylic.

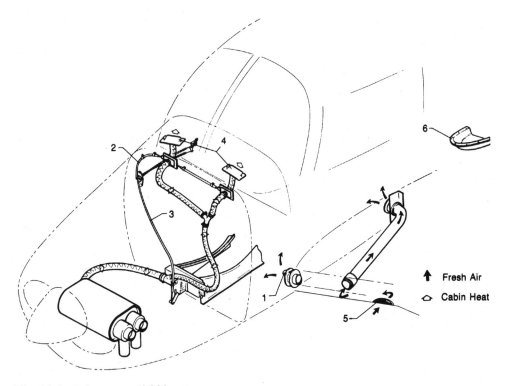

Fresh Air

Cabin Heat

Fig. 11-1. *Exhaust manifold heater.*

There is also a threat of carbon monoxide (CO) poisoning if the exhaust stack leaks inside the shroud. During preflight, it is always a good idea to check exhaust-stack seams to assure that the welds are solid; any leakage within the engine compartment may cause the cabin to fill with CO. It is impossible to check the exhaust weld within the heater shroud, though, and it provides a direct route for CO to travel into the cabin. The best safety measure is to put an inexpensive carbon monoxide detector in every airplane. When exposed to CO, the colored pad turns from orange to black in fifteen minutes or less, depending on the amount of carbon monoxide, alerting the pilot to a serious hazard.

Combustion Heater

If you think there ought to be a better way, you're right; there is. It's called a combustion heater, and most multiengine airplanes have one. But why don't singles? For the most part, airframe manufacturers feel it is too expensive to put combustion heaters in single-engine aircraft. So, at least for the time being, only multiengine pilots will have the luxury of instant heat.

Piper installs Janitrol combustion heaters exclusively in all of its twin-engine aircraft built after 1964. Janitrol shares the rest of the market with Stewart Warner's Southwind heater. The fundamentals of aircraft combustion heating haven't changed much over the years. According to the *Aircraft Heating Digest*, Volume 1, Number 1, published by Janitrol Aircraft in February, 1949, there are four main requirements: fuel for combustion, air for combustion, ignition to start combustion, and air to carry away the heat produced by combustion. That was true for the DC-3, and it's still true for modern aircraft.

Heat is produced by burning a fuel/air mixture in a heater combustion chamber. This is somewhat of a mixed blessing, because while it conveniently uses fuel drawn from the aircraft fuel tanks, it also reduces the aircraft's range when the heater is in use. Nonetheless, when you get into a cabin that is below freezing and can have near instantaneous heat without even starting an engine, a little less range—at least to this northerner—doesn't seem so bad.

The Janitrol heater uses a spray nozzle to send regulated, atomized fuel/air mixture into the heater combustion chamber. There, a high-voltage spark plug powered by the aircraft's electrical system provides continuous ignition. Because aircraft attitude and altitude always are subject to change—sometimes rapidly—Janitrol uses what it calls the "whirling flame" principle. The fuel/air mixture enters the combustion chamber tangent to the chamber's surface as depicted in Figure 11-2. This forces the airflow to spin and mix with itself, causing a stable, continuous flame pattern. The burning gases flow the length of the combustion tube, double back over the outside of the chamber, go through a crossover passage to an outer radiating area, travel down the length of the heater one more time, and finally exit through the exhaust.

The cabin ventilation is ducted separately between the combustion air chambers. Though the two airflows never mix (to do so would lead to CO poisoning), the ventilating air does contact several surfaces heated by the combustion air, causing heat transfer. Several other components round out the system. An electric fuel pump is necessary,

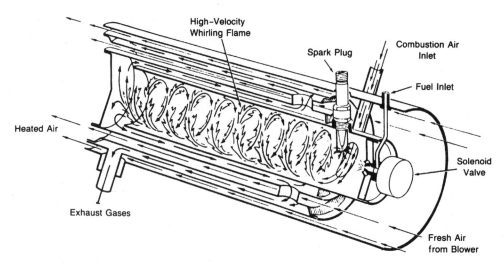

Fig. 11-2. *Janitrol heater cutaway.*

though some aircraft actually may use the engine-driven fuel pump if the fuel pressure output is correct. There must be a ventilating air blower, which also doubles as a cooling fan when the heater is turned off, and a separate combustion air blower. Temperature control is maintained by either a duct switch or cycling switch, which senses heat output and compares it to the selected temperature. And, of course, there are the requisite controls and lights that indicate the operational status of the heater.

Preflight

System preflight should include checking for either blockage or damage to both the ventilation and combustion air inlets, the heater-exhaust outlet, and the heater fuel drain. Additionally, the area around the heater exhaust tube should be checked routinely for soot accumulation; this indicates an excessively rich fuel/air mixture (which is caused by incorrect fuel pressure), a blocked combustion-air inlet line, an inoperative or failed combustion blower, or a clogged fuel nozzle.

In addition to the visual preflight, an operational check should be accomplished. First, turn the heater master switch on and assure that the ventilation and combustion blowers work; the heater-failure light also should illuminate, indicating that the system is activated, but there is no combustion. With the master switch still on, check for excessive current draw, any unusual vibrations, or noises. Then perform the operational check as outlined in your pilot's operating handbook (POH).

Troubleshooting

Six basic operational problems can occur that require troubleshooting on your part: Heater fails to light; ventilating air blower fails to run; combustion air blower fails to run; heater fires but doesn't burn steadily; heater starts, then goes out; and heater fails to shut off.

If the heater fails to light, the first consideration is procedure; double-check the POH to make sure you are using the correct one. Beyond that, insufficient electrical power, such as a dead battery or insufficient fuel, should be suspected. Some mechanical problems are beyond the control of the pilot, such as restricted fuel nozzle or inoperative fuel pump.

Failure of the ventilating blower to run probably will mean you forgot to turn the heater master switch on. Otherwise, it's a mechanic's job, as is failure of the fully automatic combustion air blower.

When the heater fires up but doesn't burn steadily, the culprit probably is fuel related—an insufficient amount or contamination by ice or water. Other mechanical problems, such as a fouled spark plug, can produce the same results.

If you suddenly realize that you can see your breath and you haven't had garlic for lunch, the heater has probably gone out. While mechanical problems could be the cause, more likely the problem is either fuel or electrical starvation. Best bet is to check the fuel supply and master switch.

Finally, if the heater fails to shut off during shutdown, it is a mechanic's problem such as a defective heater switch or stuck fuel solenoid valve.

The crux of the matter is proper maintenance. Preflight, preventive, and periodic maintenance are the keys to efficient and safe operation. Your zero-time overhauled Janitrol heater is certified to run 500 hours (or 24 months, whichever comes first) in accordance with Janitrol AD Note #96-20-07 before a pressure decay test is required. After that, every 100 hours (or 24 months, whichever comes first) another decay test is due. To avoid having to continually comply with this airworthiness directive, you can replace the unit with a new JanAero extended-life heater assembly, which consists of a ceramic coated combustion tube. Performed by a mechanic, the preventive maintenance is a thorough inspection of the entire unit, including a pressure check of the combustion chamber.

Southwind also has a maintenance-related airworthiness directive. Southwind AD #81-09-09 requires a 250-hour inspection and, at 1000 hours time in service, it must be overhauled in accordance with the manual. Incidentally, it is important to note that the "hours" referred to are actual heater-operation hours. While some aircraft have a heater hour-meter that records operating time, many do not. Janitrol allows the operator to compute one hour of heater operating time as the equivalent of two flight hours. Southwind, on the other hand, requires straight flight hour time if you don't have an hour meter on the unit itself. In general, it is probably going to be cost-effective to have an hour meter installed on all heaters.

AIR CONDITIONING

Growing up in the 1950s, a frequent sight was "It's Cool Inside" emblazoned across theater marquees. The lure of air-conditioning, something unheard of in homes, was enough to attract crowds on any hot summer night regardless of the picture that was showing. Air-conditioning spread to restaurants, to other public places, and finally to homes. Previously inured to heat, consumers quickly began to expect to be kept cool indoors; portable units found their way into cars and finally commercial aircraft. If asked, the average pilot would probably tell you it isn't practical to air-condition small aircraft. We tend to be most concerned about protecting the pilot from extreme cold, forgetting that excessive heat is also a problem.

Studies conducted by both the U.S. Air Force and the U.S. Army show that an airplane with a stable, comfortable cabin temperature is a safer flying environment. With 30-minute waits on the ground at some of the larger airports, and outside air temperatures of 80–90 degrees Fahrenheit (F), it's no wonder that the inside of an airplane can exceed 100 degrees. How safe can a pilot be after sitting in a 100-degree cabin for 30 minutes prior to takeoff? Certainly, few business executives are going to sit in that kind of heat.

There are penalties to be paid for air-conditioning, to be sure. There is an increase in aircraft empty weight, due to the compressor and other required equipment. This translates into fewer bags, reduced fuel, or fewer passengers. Just the operation of the system causes a reduction in available engine horsepower. The Cessna 210N operating handbook states that there is a one-knot TAS cruise reduction when air-conditioning is installed on the aircraft, and an additional one to two-knot TAS cruise reduction when the compressor actually is operating!

Two types of heat affect airplanes: aerodynamic and sun. Aerodynamic, also known as adiabatic skin temperature, is the result of free-stream kinetic energy being converted to thermal energy when the free stream air is slowed to zero at the surface of the airplane. The faster the airplane moves through the air, the greater the heat buildup, skin temperature being a function of free-stream temperature and Mach number. For instance, at Mach 2.0, the fuselage temperature would be approximately 260 degrees F; at Mach 5.0, it would be about 1550. This obviously is a problem for large aircraft, not singles or light twins. For the slower aircraft, the basic problem is the sun and little or no ventilation to carry off cabin heat. The automotive air-conditioner fits nicely into this type of airplane.

System Overview

Fundamentally, air-conditioning is simple physics; the rapid expansion of fluid causes a drop in temperature. There are two basic types of air-conditioning units; air cycle machines (ACM) and vapor cycle systems. Large aircraft ACMs bleed compressed air from the turbine engine and allow it to expand, causing cooling. With the vapor-cycle system used in light aircraft, a pressurized liquid refrigerant evaporates, causing a temperature reduction. This liquid refrigerant, called Freon, usually is F-21 (dichloromono-fluoroethane) or F-12 (dichlorodifluoromethane). The two, which are not interchangeable, are chosen because they are nonflammable, nontoxic, and do not cause irritation.

The system is divided into two parts as shown in Figure 11-3. They are a high-pressure side and a low-pressure side. The high side begins at the compressor discharge of high-pressure refrigerant vapor. Driven by the engine through a belt and pulley system, a clutch disconnects the compressor when cooling is not required. The low-pressure, low-temperature refrigerant vapor enters the compressor, and its pressure and temperature are raised by compression, turning it into high-pressure liquid. Then the refrigerant passes through copper coils surrounded by cooling fins to maximize refrigerant heat transfer to outside air. The condenser hangs under the fuselage in most aircraft and retracts into it

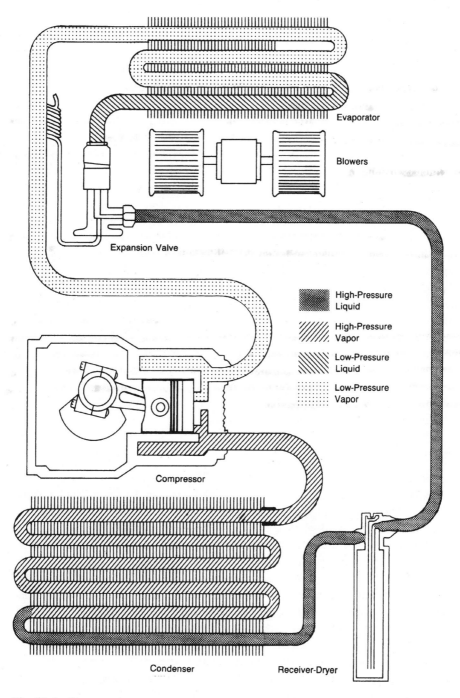

Evaporator

Blowers

Expansion Valve

High-Pressure
Liquid

High-Pressure
Vapor

Low-Pressure
Liquid

Low-Pressure
Vapor

Compressor

Condenser

Receiver-Dryer

Fig. 11-3. *Vapor cycle system.*

when the system is inoperative. Because of the excessive amount of drag the condenser causes, most aircraft have a throttle interlock switch that automatically retracts it when full power is applied, while simultaneously disengaging the compressor clutch from the engine. After leaving the condenser, the liquid refrigerant goes to the receiver/dryer.

The receiver/dryer functions as the system's reservoir. It contains a desicant (typically silica gel) that absorbs moisture; a single drop of water can freeze, lodge in the expansion valve, and completely stop the system! There also is another problem when water and refrigerant mix; they form highly corrosive hydrochloric acid, which literally will eat up the system from within. To further prevent particle blockage of the expansion valve, a filter is installed at this point. There also is a sight glass in the receiver/dryer, similar to the one in automobiles. With the system running, the sight glass should appear perfectly clear; bubbles mean flow fluid level.

Next is the thermal expansion valve. This meters the refrigerant and maintains high pressure upstream. As the liquid leaves the valve, it is sprayed through the coils, expanding as it goes, assuring complete evaporation of the liquid by the end of the coils. The valve varies refrigerant discharge, depending on the amount of heat to be removed from the cabin. This is where the low-pressure side begins, with the refrigerant turning into a low-pressure liquid. It is effectively the low-pressure equivalent of the condenser unit and consists of parallel circuits of copper tubing with fins. As the hot cabin air passes around the evaporator fins with the help of the blower, its heat transfers to the refrigerant and continues on into the cabin as cool air. This heat raises the refrigerant temperature to boiling, causing it to change state from liquid to vapor. One indication of insufficient Freon is little or no cooling air and a buildup of frost on the evaporator; a hissing sound in the evaporator is yet another indication. Completing the system, the liquid Freon now goes back to the compressor to begin the process again.

Preflight

Preflighting the air-conditioner consists of a visual inspection of the compressor, drive belt, pulley system, hoses, condenser inlet, and condensation outlet drain. The emphasis should be on system integrity—primarily damage or other signs of obvious system failure. When the system is turned on initially, it should operate within one to two minutes; otherwise, shut it down. As with all systems, it is important to read and follow the POH carefully with respect to preflight, operation, preventive, and periodic maintenance.

As adaptable as humans have proven themselves to be over the centuries, they still have a very limited temperature range. Within that range, there is an even smaller one that dictates comfort and efficiency. Properly maintained environmental systems promote safe, comfortable flying.

12
Pressurization Systems

T HE MAIN REASON FOR PRESSURIZING AN AIRCRAFT IS FLEXIBILITY. Being able to select a higher altitude may give you the option of a smoother ride, shorten your flying time, and/or provide an alternative to flying in severe weather or icing conditions. A pressurized aircraft can provide a comfortable cabin environment at significantly higher altitudes than one that is unpressurized, in which the passengers are required to wear oxygen masks.

Pressurization was originally developed in support of the WW-II effort and particularly for use in the high-altitude Boeing B-29 Superfortress bomber. Pressurization allowed the crew to move about the cockpit and passenger compartments in relative comfort while being able to take advantage of the benefits of high-altitude operation, such as more favorable winds, greater wing efficiency, and turbo-supercharged engines.

Today there are a number of light aircraft that feature pressurization systems that result in shorter flying times, lower fuel burns, higher endurance, and weather avoidance. Many pilots feel that they can reap the same benefits simply by having oxygen onboard the aircraft and breathing through individual masks, but there are some subtle physiological drawbacks to doing so.

For one, oxygen masks are both cumbersome and uncomfortable. Using a microphone with a mask is challenging. Having a simple conversation with someone else is difficult enough, as the mask muffles what you are saying. Overall, the effect is to increase both fatigue and the psychological burden on the pilot. Then consider the perspective of the nonpilot passenger who is used to flying on the airlines where they can have a drink and breathe normally in a temperature-controlled, comfortable cabin. Let's face it; who wouldn't be at least mildly uncomfortable with the idea of having to wear an oxygen mask in such a strange environment?

Cabin pressure is maintained by "packing" air at a fairly constant flow through a sonic nozzle and then controlling the flow of air out of the cabin. Because there is already a constant flow of air out of the cabin through cracks, doors, window assemblies, and other leakage points, it is far easier to pump in more air than is necessary and maintain the desired cabin pressure by regulating the opening of an outflow valve. Initially, manufacturing capability made leak-tight cabins impossible, but over the years it has become evident that developing a leak-tight cabin is not only disproportionately expensive but may not be such a good idea anyway. The constant airflow keeps cabin air fresh.

FIXED ISOBARIC SYSTEMS

In the early days of general-aviation pressurization systems, a fixed isobaric system was used that consisted of a primary valve and a secondary, or safety, valve. The primary valve utilized an aneroid that was factory preset to maintain a given cabin altitude, typically 8000 feet. The safety valve, independent of the primary, was set to open under any one of three conditions: when the cabin experienced maximum delta p, negative delta p, or when the aircraft was sitting on the ground.

Maximum delta p represents how much cabin pressure is allowable relative to the lower, outside ambient air pressure. It is analogous to how far you can blow up a balloon safely. On start-up, once the engine is operating, there is some airflow into the cabin. As power is increased to takeoff, the inflow rate is sufficient to pressurize the cabin. The aircraft leaves the ground and climbs in an unpressurized mode with the valve open. Throughout this portion of the climb the cabin pressure parallels the outside ambient air pressure until it reaches the fixed set point of the aneroid. As the aircraft climbs through 8000 feet, the valve closes and maintains an 8000-foot cabin altitude even though the aircraft continues its ascent. The aircraft maximum altitude is limited only by the structural strength of the cabin—maximum delta p. If the aircraft continues to climb, the valve will open up as necessary to prevent the cabin pressure from exceeding maximum delta p. The effect under those conditions will be for the cabin altitude to climb above 8000 feet.

Negative delta p represents a situation where the outside air pressure is greater than the pressure inside the cabin. Negative pressure is, well, a negative situation, as the airframe structure is designed to contain pressure like a balloon rather than withstand outside pressure like a submarine. Negative delta p occurs when the aircraft descends faster than the valve can outflow cabin pressure, resulting in cabin pressure greater than ambient.

The third condition, when sitting on the ground, is important because if both valves remained closed on the ground there could be sufficient pressure to make it difficult to open cabin doors and/or emergency exit windows in an emergency.

VARIABLE ISOBARIC SYSTEMS

Today's light aircraft cabin pressure systems are direct descendants of the earlier fixed isobaric systems. Instead of having a fixed system, pressurization manufacturers relocated the aneroids to a controller in the panel and ran pneumatic lines to the valves. Turning the altitude select knob on a Garrett controller simply rotates the aneroid directly, allowing more air to enter or leave the valve and causing the cabin altitude to change.

Such a direct linkage does have its problems, as it makes setting cabin pressure in flight difficult. Even a small change in dial setting can cause a rapid change because there is no rate of change control. Newer systems incorporate a rate of change function so the dial can be moved in flight, allowing for more flexibility.

Take, for instance, Janitrol's pressurization system for the Cessna P210. The system has four basic modes of operation: unpressurized, isobaric, differential, and negative relief. The unpressurized mode is in effect any time the aircraft is at a lower altitude than the cabin altitude requested by the pilot; this is common during takeoff, climb, descent, and landing. The isobaric mode begins when the aircraft climbs through the selected cabin altitude, which may range from below sea level to 10,000 feet. In the P210, the pilot selects the desired cabin altitude on the manual controller prior to takeoff; no other input is required through takeoff, climb, and level-off. If a change of aircraft cruise altitude is required, the pilot slowly adjusts the controller to preclude abrupt cabin altitude changes, which can be uncomfortable for passengers. Unless, of course, you don't like your passengers and you like that "bug-eyed" look.

The manual controller has two altitude scales, as shown in Figure 12-1. The outer scale indicates cabin altitude; the inner scale indicates the corresponding aircraft altitude at the maximum operating cabin pressure differential, which is the ratio between inside and outside air pressures. These numbers on the controller face must be multiplied by 1000 feet to determine the appropriate altitude. The pilot turns the cabin rate control knob to adjust the rate at which the cabin pressure "climbs" or "descends" to the altitude set on the manual controller. The differential pressure mode goes into operation whenever the maximum cabin-to-ambient pressure differential is reached. Because differential pressure is a measure of internal stress on the fuselage skin, if it were to become too great, structural damage to the fuselage might occur.

The transition from the isobaric mode to the differential control is automatic. The operating differential normally is maintained by the outflow valve with the safety valve acting as a backup, allowing a pressure differential only slightly higher than what is regulated by the outflow valve. The reason for the slightly different pressure differentials between valves is because if it were the same on both the primary and safety valve, the two valves would "talk" or open and close opposite of one another, which can be uncomfortable for passengers.

Maximum Pressure Differential

The maximum pressure differential value varies from aircraft to aircraft, depending on system and structural limitations and the type of operation for which the aircraft is

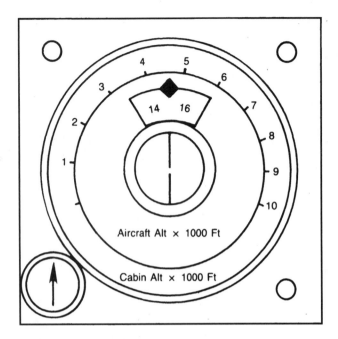

Fig. 12-1. *Cessna 340 controller face.*

designed. An aircraft such as the Beech Baron 58P, which has a maximum cabin differential pressure of 3.65 pounds per square inch (psi), is limited to a difference of 3.65 psi between the ambient air pressure and the cabin air pressure. On the ground, there is a 1:1 cabin-pressure-to-ambient-air-pressure ratio, but as the Baron 58P climbs, that ratio changes until there is a 3.65-psi pressure difference. Table 12-1 shows that the 58P is capable of maintaining sea-level pressure up to approximately 7000 feet. At that altitude the difference between sea-level pressure and the 7000-foot atmospheric pressure is 3.4 psi (14.7 psi − 11.3 psi), nearly the maximum pressure differential. At 21,000 feet and maximum cabin differential pressure, the cabin altitude would be approximately 10,000 feet.

Another factor that determines the maximum possible cabin pressure is the type of pressurization system used. The higher the aircraft is designed to operate, the greater the maximum differential needed, and the stronger the compressor output capacity required. Turbine engines can maintain high-pressure airflow into the cabin up to very high altitudes by using air from the compressor bleed-air section of the engine. Aircraft with reciprocating engines use air from the compressor section of a turbocharger similar to the system illustrated in Figure 12-2.

In light, twin-engine aircraft, powerplant failure—or even a significant, intentional power reduction—can cause the cabin altitude to rise when there is a high cabin pressure differential. This can occur because the turbocharger, which is the source of the pressurizing air, is powered by the engine. Power reductions in single-engine aircraft have the same effect, so descents should be initiated far enough in advance so the power will not have to be cut back. For the same reason, pilots should be careful not to run a fuel tank dry in a pressurized aircraft. Depending on the amount of uncontrolled cabin leakage, cabin altitude may rise faster than you can switch tanks and get the engine running again.

CABIN AIR TEMPERATURE

When air is compressed, it increases in temperature. Turbine-engine bleed air is so hot it always requires cooling before entering the cabin, even if warm air is desired. Larger turbine-powered aircraft run the pressurized bleed air through either air-conditioning packs or a vapor cycle air-conditioning system prior to cabin entry. Air from a reciprocating-engine turbocharger may require cooling only on warm days when the aircraft is flying at lower altitudes. This typically is accomplished by routing the air through a heat exchanger where the pressurized air ducting is cooled by ambient ram air. At very cold ambient temperatures, when considerable heat is required in the cabin, the pressurized air may not be warm enough and a cabin heater will be required.

Table 12-1   Standard Atmospheric Pressure

Altitude (feet)	Pressure (psi)	Altitude (feet)	Pressure (psi)
Sea level	14.7	18,000	7.3
1,000	14.2	19,000	7.0
2,000	13.7	20,000	6.8
3,000	13.2	21,000	6.5
4,000	12.7	22,000	6.2
5,000	12.2	23,000	5.9
6,000	11.8	24,000	5.7
7,000	11.3	25,000	5.5
8,000	10.9	26,000	5.2
9,000	10.5	27,000	5.0
10,000	10.1	28,000	4.8
11,000	9.7	29,000	4.6
12,000	9.3	30,000	4.4
13,000	9.0	35,000	3.6
14,000	8.6	40,000	2.7
15,000	8.3	50,000	1.7
16,000	8.0		
17,000	7.6		

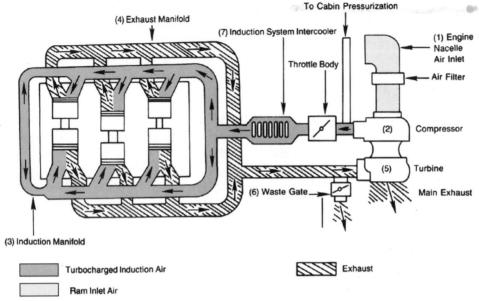

Fig. 12-2. *Cessna 421 turbosystem schematic.*

THE OUTFLOW VALVE

The outflow valve, which vents the cabin to the outside air, has three main functions: negative pressure relief, isobaric control, and differential control. Negative pressure relief is automatic, so the aircraft is never subjected to an outside air pressure greater than cabin pressure; higher pressure air always can flow freely through the outflow valve into the aircraft.

Isobaric control typically maintains cabin pressure within +0.05 psi from what the pilot selects on the manual controller. If the cabin pressure exceeds that selected on the controller, the outflow valve increases to compensate; as the pressure falls below, the opening decreases slightly.

Differential control is preset by the factory. The isobaric pressure requested by the pilot will be maintained until cabin pressure reaches the maximum pressure differential, then the differential control overrides the isobaric mode so the cabin altitude will vary directly with aircraft altitude.

If the outflow valve were to stick closed, excessive cabin pressure would build up quickly. To prevent overpressurization resulting from a stuck outflow valve, the system has a safety valve, which functions to relieve negative pressure, to provide backup differential control, and to act as a solenoid-operated cabin-pressure dump mechanism that can be operated from the flight deck or by a gear squat switch.

RAPID DEPRESSURIZATION

One of the most misunderstood aspects of cabin pressurization is rapid depressurization. Often inappropriately referred to as "explosive decompression," movies have depicted scenes where a single bullet shot through the fuselage has sent people flying about the cabin. It makes for high drama but it is pure nonsense. If you can have a fairly large outflow valve open to the atmosphere, what difference would a small bullet hole make? As long as the bullet hole remains small, the outflow valve can compensate instantly.

A pilot may depressurize the cabin intentionally if the pressurization system begins to pump contaminated air into the cabin. Also, if the pilot discovers a window is cracking, depressurization will relieve the pressure differential and take stress off the window.

There are two ways to depressurize a cabin intentionally. The gentlest would be to increase the cabin altitude slowly with the manual controller until there is no longer a pressure differential; this would be an appropriate method in the case of a cracked window, limited smoke, or fumes. In a life-threatening situation, such as a cabin fire or dense smoke, the pilot could elect to activate the depressurization switch. This reduces the pressure differential to zero rapidly, but not instantly; the outflow and safety valves are not that large!

One of the frequent causes of rapid decompression is the improper closure of a door. Unlike larger aircraft with plug doors that open into the cabin and are pushed into the fuselage by cabin pressure, light aircraft doors generally open outward, away from the fuselage. The manufacturer has to devise a locking mechanism that is easy to use but strong enough to ensure the door will not open in flight. If someone incorrectly locks the door, it could blow out when the pressure differential increases. The potential for this problem, which is almost always caused by human error, is significant enough in the King Air 200 that when I worked for FlightSafety we used to instruct all crews never to permit passengers to lock the cabin door; it was to be locked only by a crewmember. This is good advice for all pressurized aircraft. If it isn't intuitively obvious, the reason that light aircraft doors open outward is simply because of the limited room inside the cabin.

Even if a door or window did burst open in flight, it isn't very likely that passengers of light aircraft are going to be sucked out of their seats and pulled through it. An analogy would be if you filled the airplane with water and punched a hole in the fuselage; the bigger the hole, the more the water would tend to pull you toward it. Again, this is predominantly high drama in films; however, when flying as a passenger I find myself, for some inexplicable reason, never choosing a seat next to a door!

There should be concern for people who have ear or sinus trouble, however, because instantaneous pressure changes can be painful. Cold, fear, and lack of oxygen present a far more serious problem than the rapid decompression itself. It feels as if someone has stepped on your chest as the air suddenly expands in your lungs, forcing its way out through your nose and mouth. You couldn't hold your breath if you tried, but it would never occur to you to try. The cabin develops a condensation cloud for a short time, making it difficult, if not impossible, to see within the cabin. This could be a short-duration problem for the pilot, as instruments could become difficult to see. Very quickly after decompression it gets incredibly cold, which becomes a major concern. Panic could be a problem for passengers with heart trouble.

Losing cabin pressure when flying above 10,000 feet probably will warrant immediate descent. Even with supplemental oxygen available, the cold can be deadly. The best course of action in that event is as follows. First, fly the aircraft! Nothing else matters if you lose control of the aircraft. Second, don an oxygen mask and make sure you have oxygen flowing. Third, pause for just a second to shake off the fear, then check the passengers to make sure they are on oxygen. Finally, begin a descent while assessing the situation and determine the best course of action based on cabin temperature, oxygen available, structural condition of the aircraft, weather, distance to the airport, wind conditions at a lower altitude, and the condition of the passengers. Don't underestimate the physiological effect of high-altitude, unpressurized operation on you or your passengers. Your best insurance policy in this type of situation is to have taken the time in advance to be familiar with applicable emergency procedures in the pilot's operating handbook. Also, be aware that a lower altitude frequently means increased turbulence, so descent should be at a reasonable speed, not at V_{ne}.

Another concern about rapid descent is existing structural damage because the airframe may not hold up under high-speed descent. Also remember that as you descend, air density increases and so will the indicated airspeed for a given deck angle.

There are other concerns related to cabin depressurization. If sitting in the freezing cold contemplating a course of action sounds like a great time for a cup of hot coffee, remember that the thermos was sealed at ground pressure, so it is a potential bomb. Don't forget to squawk 7700 on the transponder. If you are in instrument conditions and it becomes necessary to use alternate instrument air, be aware that a large hole in the fuselage can cause the cabin pressure to be lower than ambient due to a venturi effect. Perhaps most often overlooked is a passenger briefing before the flight; a little knowledge can go a long way, especially concerning the use of oxygen and the effects of smoking.

PREFLIGHT AND OPERATIONAL CONSIDERATIONS

During the preflight, make sure the door is properly sealed and the dump switch is off. After engine start, to assure the system will work while still on the ground, set the aircraft altitude controller to 500 feet below field elevation. Now pull the landing gear circuit breaker and increase the rate controller; the system should begin to pressurize the cabin because you have overridden the gear squat switch and tricked it into "thinking" it was flying above the selected altitude. Then test the dump switch to make sure it will work if you should need it in flight. You should never take off in a pressurized condition because the aircraft is not designed for it.

In preparation for takeoff in the Cessna 340, which uses the Garrett AiResearch system, the procedure is somewhat different than in the Cessna 210. AiResearch instructs pilots to select 500 feet above field elevation on the cabin altitude selector and set the cabin rate control knob to the 12 o'clock position. Then start the engines and check for airflow into the cabin to assure it will pressurize after takeoff. There are two reasons for doing this: First, it prevents the pressure "bump" sometimes felt on takeoff as a result of both the safety and outflow valves closing simultaneously. The safety valve, which closes when the gear retracts, is controlled by the squat switch. The outflow valve closes when the cabin reaches the altitude you have requested. If the controller is set to field elevation, both may slam shut simultaneously on takeoff. With the controller set to 500 feet above field elevation, the outflow valve will close long after the safety valve and the passengers will experience a smoother transition. The second reason is that if you set the selector to cruise altitude, the system will "prerate," meaning it will think the aircraft is climbing long before it actually does. This will delay normal cabin pressurization longer than necessary and may prove uncomfortable for some passengers.

Once the climb is established and you have passed through 500 feet AGL, reset the aircraft altitude selector to 1000 feet above cruise altitude. As the aircraft climbs, the cabin altitude takes care of itself. The reason for setting 1000 feet above cruise altitude is again passenger comfort. If the controller is set to cruise altitude, the outflow valve will open and close continuously as the cabin pressure makes small fluctuations between too low and too high. With the controller set for an altitude above the actual aircraft altitude, the cabin will never reach the programmed pressure, so the outflow valve will remain at least slightly opened; the result is no bumps. There are no additional requirements for cruise condition. If it is necessary to change altitude, simply select the new altitude plus 1000 feet and climb or descend.

During descent for landing, set the aircraft altitude selector to approximately 500 feet above field elevation and adjust the cabin rate of change to maintain a comfortable cabin rate of descent. It is a good idea to not descend at a rate that will allow the aircraft to catch up to the cabin altitude, otherwise the cabin will depressurize. On the other hand, if you select the field elevation for the cabin altitude, when the gear touches down on the runway the cabin will dump, causing some passenger discomfort. With 500 feet above airport elevation selected, the cabin will depressurize comfortably shortly before landing.

TROUBLESHOOTING

Fortunately, pressurization systems are basically reliable, but things can go wrong. Here are a few thoughts on troubleshooting.

If there is a "bump" felt at rotation on takeoff, an ear-popping event caused by a sudden pressurization, it is probably the result of the outflow valves closing too rapidly, fuselage flex, or a change in airflow over the outlet holes. Normal fluctuations in pressure, depending upon their frequency and magnitude, may not be felt by the occupants, but bumps are usually uncomfortable and can potentially cause structural damage.

If the cabin follows the aircraft's altitude and rate-of-change shown on the flight instruments, there may be several causes. The first thing to check is the dump switch. If the dump switch is off, the logical choice would be a problem in the landing gear solenoid valve, which is responsible for keeping the safety valve open during ground operations. Try cycling the gear to see if that helps, then try opening the landing gear circuit breaker to bypass the system.

If the down rate is faster than the up rate, but everything else works normally, just make the necessary adjustment manually and have the controller checked out at your next opportunity. The problem probably is a minor leak in the tubing or controller.

Should the cabin rate exceed the selected rate value during the aircraft's climb to cruise altitude, increase the rate selection or decrease the aircraft's rate of climb. The obvious answer is that the aircraft is climbing faster than the controller and it is at the maximum differential; however, it also could be a controller malfunction.

If the cabin altitude exceeds what you have selected, one of several problems may exist. There may be a loss of pressurizing airflow for some reason. The aircraft altitude may have exceeded the positive differential pressure value. There may be an internal malfunction of the controller, the outflow valve, or safety valve. Or you may have something as simple as a leak in the tubing. In any event, your only choices would be either to adjust to a higher cabin selection, if possible, or to reduce the aircraft altitude. Other problems that may arise are usually beyond the pilot's ability to correct in flight.

13
De-icing and Anti-icing Systems

ICE IS THE NEMESIS OF EVERY IFR PILOT. VIRTUALLY ALL ASPECTS OF IT are negative: It's capricious; it's fickle; should you climb, descend, turn back? The stakes are high, particularly for those who fly IFR in aircraft not certified for flight into known icing. A chance encounter with icing can produce drastic aerodynamic changes, propeller imbalance and vibration, increased drag, increased weight, and reduced airspeed. To fight back, general-aviation pilots can arm their aircraft with de-icing equipment.

It is crucial to understand that even though light general-aviation aircraft may be equipped with de-icing and anti-icing equipment, they are not intended to be flown for extended periods in known icing conditions. Rather, the intent of these systems is to give the pilot some options in situations where icing exists—for instance, an IFR descent through a layer of icing on an instrument approach. They are not designed to allow you to fly in icing conditions from Chicago to Los Angeles.

DE-ICING VERSUS ANTI-ICING

Simple as it may be, some folks don't understand the fundamental difference between de-icing and anti-icing equipment. De-icing, as the name implies, removes accumulated ice from the leading edges of wings, horizontal stabilizers, vertical stabilizers, and propellers. Yes, props have de-icing, not anti-icing, equipment!

Anti-icing equipment is used where no amount of ice can be tolerated; it requires a significant current draw and, except for the relatively small pitot tube heat, is seldom found in light aircraft. A common location of anti-icing is at turbine-engine inlets, where a chunk of ice could produce disastrous results.

The first working airfoil de-icer was designed in 1929 by William C. Geer in conjunction with the Guggenheim Safety Foundation and the National Advisory Committee for Aeronautics. His design was implemented by BFGoodrich in 1932, and the first pneumatic de-icers to be installed on a commercial aircraft were on a 1930 Northrop Alpha mail plane. A close relative of those early pneumatic de-icers is still used on the leading edges of airfoils. Since then, however, electrothermal de-icers have been added to propellers.

ELECTROTHERMAL PROPELLER DE-ICING

The colder the outside air temperature, the greater the tendency for ice to adhere to a surface. Fundamentally, electrothermal propeller de-icing is a simple matter of converting electrical energy into heat and transferring the heat to the prop. That takes a fair amount of electrical power, and when you are flying IFR in icing conditions you don't have a lot of power to spare. To solve that problem, the job is broken into two elements: outboard and inboard.

In a single-engine airplane, first the outboard element will heat up, then the inboard. This cycling—outboard, inboard—continues as long as the propeller de-ice switch is on. With a 14-volt system, you can anticipate a 20 to 23-amp draw with a two-blade prop, 30 to 34 amps with three blades, as compared to 8 to 12 and 14 to 18 amps, respectively, for a 28-volt system.

When de-icing, there is more at work than just heat. Centrifugal force is pulling constantly, especially on the outboard section. Here, a little ice buildup helps by increasing the mass of the ice; the pause in outboard heating while the inboard element is activated allows some buildup. Then, as the outboard heat turns on again, the adhesion of the ice reduces, centrifugal force tugs away, and the ice flies off into the blast of air.

The dual-element system, though not the only type, still is the most common. In the single-engine airplane, regardless of the number of propeller blades, there are two independent circuits, as shown in Figure 13-1. All outboard elements simultaneously heat for 34 seconds, then all inboard elements for 34 seconds. If all outboard elements didn't heat simultaneously, there would be a strong tendency for rotational imbalance to occur as one prop outboard section shed ice and the other didn't.

BFGoodrich offers a slightly different method of handling the situation with their HOTPROP® de-icer. This single, graduated heat element solves the problem of excessive

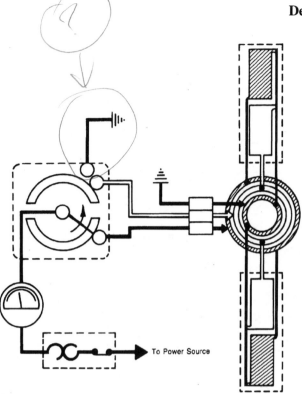

Fig. 13-1. *Dual-element de-icing system—single-engine aircraft.*

electrical load by providing high heat at the inboard end and low heat at the outboard end. The HOTPROP® de-icer is still two cycle: 34 seconds of heat over the entire length of all blades and 34 seconds of no heat.

Twin-engine aircraft compound the power demand by adding another set of prop blades. Dual element systems still are the most common, but, as shown in Figure 13-2, the cycling is different. The sequence of these 34-second cycles is as follows: right outboard, right inboard, left outboard, left inboard. Rotational balance is assured the same way as with the single-engine aircraft system.

It is important to note that the protocol for the timer in the twin-engine aircraft system varies among aircraft. They do not necessarily reset to a "home" position when deactivated. Some do, some step, and some do neither one. You need to observe your system to understand which elements will heat up first, but the sequence will always remain the same.

One of the eeriest situations associated with flying occurs on a dark night, in between cloud layers, in icing conditions. Just the situation is enough to make the average pilot squirm, but then you turn on the prop de-ice. In a minute or so there is this horrendous WHUMP as the ice flying off the prop slams against the side of the fuselage. No matter how many times you experience that situation, that first WHUMP just about makes you jump out of your seat.

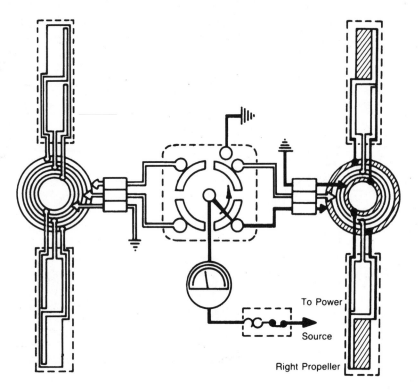

Fig. 13-2. *Dual-element de-icing system—twin-engine aircraft.*

System Components

An electrothermal propeller de-ice system can operate on either a 14 or 28-VDC system. HOTPROP(R) de-icer systems have etched foil, but with other, dual element systems, the actual de-icer elements are made of hand-wound conductive wiring embedded in a thin fabric and rubber sheet that is bonded to the first third of the leading edge of the propeller blades. The principle is simple, but putting the principle into practice is what causes trouble. The problem is how do you get electrical power out to a rotating propeller? The solution—and weak link—is a slip-ring and brush-block assembly.

This copper distribution center transfers electric current to the rotating propeller de-icers from spring-loaded carbon brushes. The brushes, fixed to the engine, maintain constant contact with the slip ring as it rotates with the propeller. Potential problems range from uneven wear to loss of conductivity as a result of oil, grease, and carbon covering the slip ring. A common pilot-induced problem results if the system is turned on and left operating for more than approximately 5 minutes when the engine is not running in ambient temperatures higher than 100 degrees F. It will most likely burn up the propeller de-icer.

With a manual system there is no timer, so the pilot is required to keep flipping the switch to activate the system. The timer in an automatic system continuously cycles the elements at 34-second intervals. Most systems also have an ammeter, which indicates the operating current. Other than a momentary fluctuation as the timer cycles to another element, the ammeter needle should always be in the shaded area when the system is activated. Prior to engine start, a low battery may cause the needle to be slightly below the shaded area. The ammeters in some systems won't show any flicker at all, and some systems don't even have an ammeter, so you need to understand the telltale signs for your own system. Rounding out the system is the circuit breaker, on/off switch, and the wiring harness.

Preflight

Propeller de-ice should be checked prior to every flight during winter months and, provided your aircraft is certified for it, whenever you are going to fly into known icing conditions. The rest of the time you should preflight check it at least once a week. The prudent pilot will follow what the aircraft's pilot's operating handbook (POH) says about preflighting propeller de-icers. While there are differences among aircraft, here are some general items to check:

Visually inspect the rubber elements carefully. Look for wrinkles, debonding, rips, impact damage, cracks, or erosion of the rubber surface. Several leading-edge tapes designed to reduce erosion are available. If they are specifically approved by the propeller or aircraft manufacturer, that's fine; they've been tested and proven effective. Otherwise, do not use them. There has been research conducted on several brands of these tapes and they have been found, in some cases, to have insufficient thermal conductivity, resulting in a lower surface temperature than may be required to shed ice!

After a good visual inspection, activate the system and put your hands on the elements. You should feel them get warm in a few seconds; if not, there is a problem. On the opposite end of the spectrum, if you put small barbecue grill marks on your hand, I'd check to see if there was a little amperage output problem with the GPU. In either case, turn off the system to prevent burning out the de-icer. In fact, prolonged use also will cause significant battery drain, and the elements can do serious damage to nonmetal propeller blades.

Finally, with some systems it is possible to conduct an operational check and see an indication on the ammeter. It is important to know what to expect from your system, as not all systems react the same way. With systems that do show an indication on an ammeter, with the engine running, turn on the de-icing system and observe the ammeter. Every 34 seconds, the needle should deflect momentarily, indicating that the timer is cycling and putting out power. It is important to understand that this doesn't necessarily mean the elements are getting the power; it only means that the timer is turning it on.

System Maintenance

With all of that slapping, scraping, rotating, and vibrating going on, there is more than just a chance things won't work exactly right. To head off trouble, you should have your mechanic do both 50-hour and 100-hour inspections; it's the cheapest insurance you can get.

The 50-hour inspection determines whether all of the current actually is getting to all of the de-icing elements at the right time. The mechanic will inspect the wiring harness carefully, measure current flow, and assure proper sequence. The 100-hour inspection duplicates the 50-hour and adds a thorough check and cleaning of the slip ring and carbon brushes.

Troubleshooting

Troubleshooting the de-icing system is a relatively simple task; you either get the WHUMP or you don't. There is very little you can do once airborne, and the ammeter is the key to virtually everything. If it reads zero, check the circuit breaker; on the ground, with the engine not running, it also may be the battery master. If the ammeter reads normal during part of the cycle and zero during another part, you've got trouble. With partial de-ice capability, rotational imbalance is likely to occur, which could be severe enough to cause structural failure of the prop. Deactivate the system.

Normal reading during part of the cycle and low current during the other part probably means that inner and outer elements are heating simultaneously. That produces a hefty current draw, but probably nothing more serious. If the ammeter always reads low, you have low system voltage, an indication of generator or voltage regulator problems. A constant high reading indicates a de-icer power lead shorted to ground, warranting a system shutdown.

In systems where the ammeter flickers, if it flicks more frequently than every 34 seconds, there probably is a loose connection. If it flicks less frequently, it's an indication of an inoperative timer and incomplete de-icing. The potential, again, is rotational imbalance. If the propeller isn't shedding ice at all, there could be a short in the wiring harness, worn out carbon brushes, even gas or oil on the slip ring and brushes. If there is oil on the slip ring, have your prop seal checked; it could be worn out.

Radio static appearing only when propeller de-ice is turned on probably indicates arcing brushes, loose connections, or a wiring harness that's too close to the radio equipment. If you can't stand the popping for another flight, turn this aeronautical Gordian knot over to your mechanic and make sure you set a maximum dollar amount allowable in the effort to unravel it!

PNEUMATIC DE-ICING SYSTEMS

If ice will adhere readily to a propeller, it will collect massively on a wing or horizontal stabilizer. Because the wing is so large, electrothermal heating really isn't practical. While more sophisticated aircraft have weeping wings, hot wings, and other expensive equipment, most of us live with pneumatic de-icing systems that we affectionately call "boots."

Here, the principle and practice are relatively simple. As ice accumulates on the leading edge, you mechanically expand it and break the ice loose. This requires inflatable rubber de-ice boots, a pneumatic system, timer, and relay switches for inflation sequencing, an on/off switch, and a pressure gauge or indicator. Actually, it is more straightforward and simple, as a look at the schematic in Figure 13-3 will illustrate. The heart of the pneumatic system is the pump. Its pressure side inflates the boots and its vacuum side deflates and holds them down. Other components include tubing, which seems to run endlessly through the wings and fuselage, flow control valves to channel airflow to the boots, regulators to control both pressure and vacuum level, and pressure-relief valves.

The boots are fabric-reinforced rubber sheeting bonded and stitched so as to produce parallel, inflatable tubes that may run along the span or the chord. All rows may inflate simultaneously, or they may inflate alternately, but all systems are essentially the same. If you are not experienced at flying in icing conditions, be warned that some dual instruction is necessary before tackling it alone. Handling icing is not as easy as turning on the boots and autopilot. You need to let about a half-inch of ice build up before activating the boots for maximum effectiveness.

For many years there was a concern that activating the boots too often would cause "bridging," a condition where the ice would build up over the extended boots and then become impossible to break up. According to BFGoodrich there is no test data or in-field reports substantiating that concern. Their experience has been that it may take the de-icer more cycles to remove ice, because not enough buildup has occurred, but the ice eventually builds up sufficiently and is removed by cycling.

Preflight

Preflight the boots by visually inspecting them for cracks, rips, tears, bubbles, holes, or separation from the airframe. You are looking for the general condition of the rubber. During the cockpit check, with the engines at runup power and brakes locked, activate the boots. If your system is so equipped, watch the vacuum pressure gauge as the boots inflate. There will be a momentary drop in pressure, but the needle should always stay

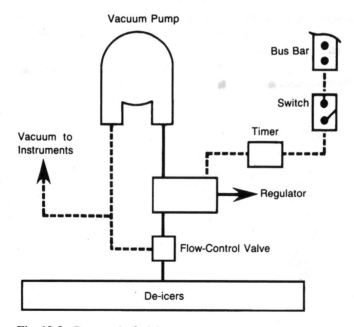

Fig. 13-3. *Pneumatic de-icing system.*

within the green arc. If it drops below the green arc either you aren't carrying runup power on the engines or there's a leak in the system or holes in the boots. If all is well, visually observe the boots inflating and deflating in the proper sequence.

Some systems operate wing, horizontal stabilizer, and vertical stabilizer boots simultaneously, while other systems sequence them. Check your POH and assure that the system is sequencing correctly. Inflation cycles generally take about six seconds, so let them cycle at least three times. This is also a good time to check for softballing—a ballooning effect that has three possible causes.

The most common cause of softballing is debonding, where the de-icer comes unglued from the airfoil over time, generally the result of poor installation technique. With a sewn de-icer, a broken stitch can cause ballooning and is often the result of a lightning strike or FOD damage but can also be the result of aging and normal wear. The third possibility is delamination, which is the separation of the outer neoprene layer from the undersurface, as seen in Figure 13-4. Delamination is more likely in older boots, as new materials and process improvements established since the early 1990s have effectively eliminated that problem.

Troubleshooting

Boots also may have one of those Gordian knot problems. During preflight, everything checks out, but at altitude some of the boots don't inflate. Indignantly, you land, taxi up to the shop, and curse silently as the mechanic tells you there's nothing wrong. The problem is subtle but not really difficult to understand and locate. You almost certainly have pinholes in the boot, and they're probably caused by erosion, FOD, or simply ozone deterioration. These tiny holes are too small to prevent proper inflation on the ground, but in flight, when flying through rain or clouds, the vacuum that holds the boot flat will draw moisture through them. Because few pneumatic systems have water separators, the moisture collects in the valves, lines, and boots. At altitude, the moisture freezes and prevents pressure from reaching the boot, then unfreezes when you descend. Those same

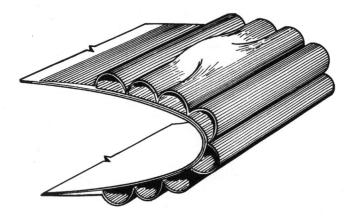

Fig. 13-4. *Ballooning effect of de-icing boots.*

holes can also allow ram air in flight to slightly expand the boots, which may appear as if they are inflated when they really aren't. On the ground, absent the ram air, they lay normally.

If a boot never inflates, the first thing to look for is a hole. Activate the system and liberally apply a 50% soap-and-water solution, which should quickly locate the problem. If the bubble routine doesn't work, you have a more serious problem and it's time to call your mechanic. The culprit could be dirty contacts preventing the flow valve from getting the go-ahead signal. More likely, it is a dirty valve, and removing, partially disassembling, cleaning, lubricating, and reassembling it will do the trick. Other possible problems include leaking or broken lines, timers, or relays.

Boots that only partially inflate probably indicate an incorrect pressure-regulator setting, but as the regulator is factory preset, it shouldn't be adjusted in the field. If that's not the problem, then the valves could be stuck in a less than full-open position, the lines could be clogged, or the pump could be simply going bad. The easy one is when boots inflate out of sequence; that's timer trouble. On the other hand, boots that inflate when the system is off require some thinking.

When in flight, the airflow over the wing causes a partial pressure around the boots. That is why a vacuum is applied to the boots when they aren't inflated—to prevent the boots from falsely inflating due to the external partial pressure. If there is a leak in the vacuum lines the ambient partial pressure will pull the boots away from the airfoil in a sort of false inflation. If they automatically inflate in flight with the system off, you have a potentially serious condition because even partially inflated boots can erode aerodynamic efficiency, making takeoffs and landings very dicey indeed.

Preventive Maintenance

Preventive maintenance of pneumatic de-icing systems will go a long way toward extending their life. Your mechanic can do a lot for holes and tears with a cold patch kit. If done properly, a patch will last the lifetime of the boot, but don't expect miracles. The process is similar to patching a tire, but you must use a patch kit designed for de-icing systems. The possibility of successfully patching a hole is determined by the size and location of the hole. For instance, according to the BFGoodrich *Installation, Maintenance and Repair Manual for Pneumatic De-icers*, the largest allowable patch repair is 5 inches by 10 inches. It also stipulates that a de-icer must be replaced if there are cuts, tears, or ruptures that cut the inflatable tube fabric, if there is broken stitch or thread, or if the damage that leaks air is larger than a 4-inch-by-9-inch area.

As boots accumulate small surface nicks and scuff damage, you may want to have the surface cover refurbished. This is a sort of "magical" cleaning up that actually will resurface the boot, provided the damage is not deeper than 0.010 inch and there is no air leakage when inflated. It is a protective coating that may be applied more than once; however, the manufacturer recommends not doing it more than twice. If the scuffing isn't too bad, it's an economical way of cleaning up the surface, and it has the added benefit of greatly improving the appearance.

Chapter Thirteen

A final note on preventive maintenance that applies to propellers, boots, and in fact all rubber parts. The ozone in the air causes premature aging of all natural and synthetic rubber. Sure signs of ozone damage are pinholes, cracks, crazing, and hardening. It is money in the bank to coat all rubber parts with Age-Master No. 1 every 150 flight hours or at least twice a year. This is probably the best protection against ozone damage currently available.

For de-icing surfaces, follow up that treatment with an application of ICEX II if it is the icing season. It reduces ice's ability to adhere to the rubber surface, making the de-icing system more efficient, but it does not maintain or preserve the boots in any way. The manufacturer recommends applying it every 50 hours to boots and 15 hours to propellers. If you are going to do it after an application of Age-Master No.1, hold off a minimum of 24 hours for curing. You can occupy those hours by using one of the few positive aspects of ice—cooling your favorite libation as you contemplate how to keep the negative kind off your airplane.

THE TKS WEEPING WING

Weeping is more than creeping into general aviation, and while it is a highly effective method of de-icing and anti-icing, it is still an expensive option. TKS ice protection offers a significantly high level of ice protection. It has the major advantage of providing anti-ice capability, as opposed to de-ice. The end result is an ice protection system that keeps ice off the aircraft while maintaining aircraft performance in the icing environment, unlike the inherent aerodynamic problems associated with pneumatic de-icing boots. This level of protection, coupled with the ease of use of the system, provides effective, simple ice protection. TKS ice protection systems have been extensively tested by NASA and have been standard equipment on the Hawker BAe-125 business jet for over 25 years.

The TKS ice protection method is based upon the freezing point depressant concept. An antifreeze solution is pumped from panels mounted on the leading edges of the wings and horizontal and vertical stabilizers. The solution mixes with the supercooled water in the cloud, depresses its freezing point, and allows the mixture to flow off of the aircraft without freezing.

The system is designed to anti-ice, but it is also capable of de-icing an aircraft. When ice has accumulated on the leading edges, the antifreeze solution will chemically break down the bond between the ice and airframe, allowing the aerodynamic forces on the ice to carry it away. This capability allows the system to clear the airframe of accumulated ice before transitioning to anti-ice protection.

A valuable side effect of TKS ice protection is the reduction of runback icing on the wings and tail. Once fluid departs the panel on the leading edge of the surface, it flows aft over the upper and lower surfaces and departs the aircraft at the trailing edge. This runback effect keeps ice accumulation in check aft of the panels from runback or from impact of larger water droplets. This side effect is a positive benefit in today's environment of concern for ice protection during large droplet encounters.

TKS ice protection systems have been developed for a number of aircraft around the world, with a majority of the recent developments occurring here in the United States. Systems have been developed for several general-aviation Beech, Mooney, Cessna, Aero Commander, and Socata aircraft, ranging from safety of flight installation supplemental type certificates for inadvertent icing encounters, up to full-known icing certifications. Mooney aircraft were the first single-engine aircraft to receive known-ice certification.

The leading edges of the wings, horizontal stabilizers, and vertical stabilizer are outfitted with laser titanium panels in which there are laser-drilled holes. The prop is fitted with a slinger ring, and a spray bar is attached to the windshield, as illustrated in Figure 13-5. All emit a glycol-based fluid that is metered from a tank, through micro-filters to proportioning units, by an electric pump. For systems that are not certified for

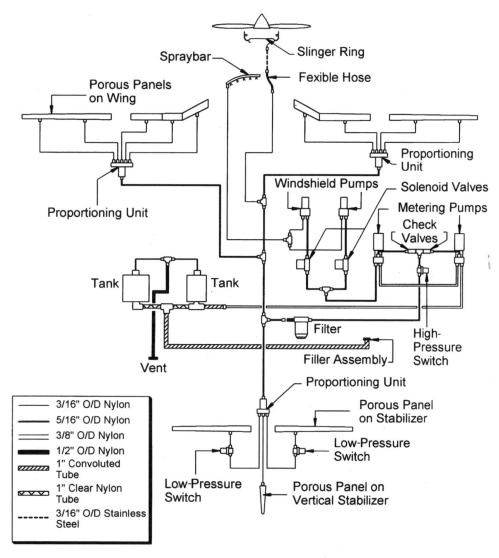

Fig. 13-5. *Flight into known icing certified TKS Ice Protection System for Mooney M20K, M20M, and M20R by Aerospace Systems and Technologies, Inc.*

flight into known icing, one metering pump is provided. For systems certified for flight into known icing, two pumps are installed for redundancy. The pumps are individually selectable.

As far as the pilot is concerned, when entering icing conditions you just turn the system on. It serves as both a de-icing and an anti-icing system. The TKS anti-icing/de-icing fluid prevents ice buildup, forms a protective film on ice-free surfaces, and chemically breaks the bond of accumulated ice without harming the aircraft's paint or other surfaces.

14
Hydraulic Systems

THE SCIENCE OF HYDRAULICS DATES BACK TO THE SEVENTEENTH century and the French mathematician Blaise Pascal. He noted that the pressure of a static liquid at any given point is the same in every direction and exerts equal force on equal areas. Figure 14-1 illustrates this principle by showing that an incompressible fluid can transmit a force, and more importantly multiply it, anywhere throughout the system.

Most pilots are familiar with the basic hydraulic system that operates aircraft brakes. It is simple, straightforward, and easy to understand, but don't let it fool you. Hydraulic systems can be as beguiling and inscrutable as the most complex electrical systems, and while they won't bite you if you touch the wrong place, they have a nasty habit of making a mess, defying common sense, and going awry at the absolute worst possible time.

OPEN VERSUS CLOSED SYSTEMS

There are two kinds of hydraulic systems: open and closed. Windmills and waterwheels are examples of open systems. The fluid passes by, or through, them and is not otherwise restricted. The force of the fluid causes the windmill or waterwheel to rotate which in turn performs some specific task while the fluid goes on its way.

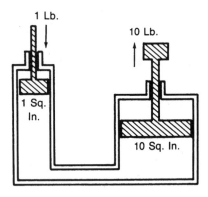

Fig. 14-1. *The pressure of a static liquid at any one point is the same in every direction and exerts an equal force in all directions.*

An example of a closed hydraulic system is the car rack at an automobile repair shop. The fluid is contained within the unit and used over and over. The hydraulic system in an aircraft is a closed system because it confines the fluid. In a closed system, fluid pressure may be increased, which increases the amount of work derived from a given amount of fluid.

There are many possible uses for hydraulics in an airplane. Some of them can alternately be accomplished with pneumatics and electric motors but overall hydraulics tends to be the system of choice for doing big jobs. The more obvious uses of hydraulics are brakes, flaps, and retractable landing gear such as shown in Figure 14-2, but there are others. Hydraulics has also been used with gear struts, engine valve lifters, shock absorbers, nosewheel shimmy dampers, antiskid systems, and control surface actuators.

SYSTEM COMPONENTS

While hydraulic systems may be used in many different ways, all systems share common components. Hydraulic systems are composed of six basic elements: hydraulic fluid, reservoir, pump, pressure control valve, fluid control valves, and actuators. All systems share these common components, while more sophisticated systems may include additional components such as an accumulator.

Fluid

Fluids are flexible. They readily change shape to fit their surroundings, can be divided into parts to work in different places, have the capability to move rapidly in one place and slowly in another, and can transmit a force in any direction.

There are two categories of fluid: compressible and incompressible. Compressible fluids, called gases, are such fluids as air and nitrogen. The branch of mechanics that deals with the properties of gases is called *pneumatics*. Incompressible fluids (liquids) include water, oil, and hydraulic fluids, to name a few. The science that deals with the transmission of energy, and the effect of the flow of liquids, is called *hydraulics*.

Hydraulic fluid is the lifeblood of the system. There are many other incompressible fluids that could transmit force, but hydraulic fluid is very special because its characteristics

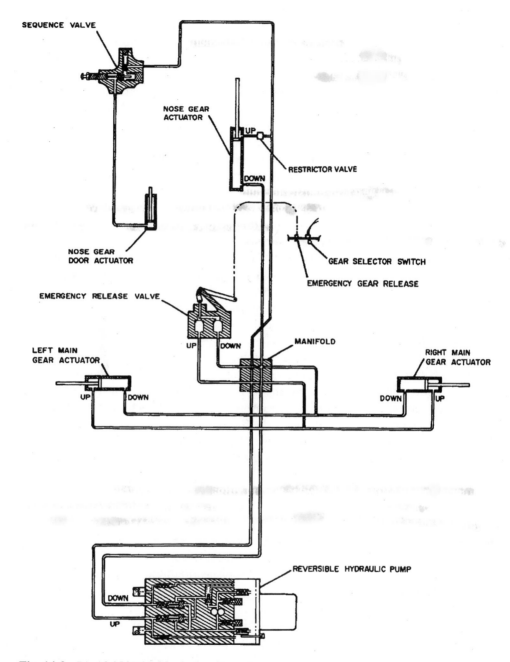

Fig. 14-2. *PA-46-350P Malibu hydraulic system.* (Courtesy of The New Piper Aircraft Company)

allow it to perform a number of critical tasks necessary for system effectiveness and long-term health. Foremost, it is an energy transmission medium, but it is much more. Hydraulic fluids also act as a lubricant for the internal moving parts of the system, and they are sufficiently viscous to seal the clearances between them.

There is one significant difference between static and moving liquids. A moving liquid is affected by friction, which causes some of its energy to be lost as heat. This problem can be minimized in designing a system if you know the five major causes of friction: tubing that is too long, too sharp a bend in the tubing, too many fittings, too high a fluid velocity, and insufficient tube diameter. In the best possible of systems, however, heat is still a natural by-product of the work a hydraulic system performs. A major function of the fluid, therefore, is that it acts as a heat-transfer medium to help cool the system.

Finally, hydraulic fluid also resists chemical breakdown, is noncorrosive, is compatible with the synthetic seals in the system, prevents rust and corrosion, and resists foaming, which prevents air from entering the lines.

In the early days of aviation, hydraulics was limited to a simple brake system using castor oil and natural rubber seals. This simple system lasted a number of years until larger aircraft were built requiring retractable landing gear, flaps, and other movable systems. The demand for higher pressure exceeded the capability of simple castor oil, so manufacturers went to petroleum-based fluids with neoprene seals, which were used extensively on WW-II aircraft.

In 1946, a single accident changed the future of hydraulics dramatically. A United Airlines DC-6 flying from San Francisco to Denver experienced a hydraulic fire that rapidly destroyed the wing. The crew attempted to land the aircraft in Bryce Canyon, Utah, but the aircraft crashed short of the runway. The aftermath of that accident resulted in the development of fire-resistant hydraulic fluids.

Today, larger aircraft use phosphate ester-based nonflammable fluids such as Skydrol, but their expense still puts them outside the range of most light aircraft. The phosphate ester fluids are approximately seven times as expensive as mineral-based fluids. So the mineral-based MIL-H-5606 fluid is used in virtually all light aircraft and can be identified by its red dye. This type of hydraulic fluid is used with neoprene seals and hoses, so no other fluid should be used in its place, as they will cause seal and hose deterioration. But the continuing and significant drawback is flammability. Great caution must be taken while pouring and storing mineral-based fluids. Having said that, there really isn't any other reason for light aircraft to use the phosphate ester fluids. They are specifically designed to operate at very high loads and temperatures, neither of which occurs in light aircraft.

At best, hydraulic fluid is touchy stuff. Make sure the reservoir cap is fastened securely to prevent leakage, and always keep fresh fluid in the system. Old fluid smells sour and the color darkens. Drain old fluid, flush the system with Varsol or Stoddard solvent, and replace it with fresh fluid.

Hydraulic systems are very intolerant of contamination. In fact, the synthetic fluid systems are so intolerant that the fluid has to be inspected with a microscope. Whenever a hydraulic system component fails, the prudent owner will drain the fluid, flush the system, and fill it with fresh fluid. It's not worth the risk of running contaminated fluid through the system only to have it eat the seals and grind the surfaces.

Reservoir

The system reservoir is basically a permanently installed can of hydraulic fluid that includes a reserve supply. Because hydraulic fluid expands as the temperature increases, the can is oversized to accommodate expansion. It also serves to release air from the fluid to prevent the air from entering the lines. Pressurized reservoirs are primarily used on aircraft that operate at high altitudes because the reduced ambient pressure causes the fluid to foam. Light aircraft, because they operate at lower altitudes and lower system pressures, use the simpler, nonpressurized reservoir.

Pump

The heart of the hydraulic system is the pump. There are a number of methods employed to power a pump, including electricity, bleed air from a turbine engine, directly off a reciprocating engine, or even by hand. While the use of hand-driven pumps in aviation has become relatively rare, some aircraft still use them fairly effectively. The Mooney Mark 21, for instance, uses a single-action hand pump for its flap system; each downstroke pumps the flaps down a little more. There also are double-action pumps, where each stroke up or down moves the flaps.

In light aircraft, virtually all major hydraulic systems are engine-driven. The important thing to remember about pumps is that they don't create pressure; they just move fluid. Pressure is created by a resistance to the movement of the fluid.

Pressure Control Valves

There are two basic methods of controlling system pressure: the pressure regulator valve and the pressure relief valve.

The pressure regulator valve, which is controlled by spring tension, senses the system pressure downstream from the pump. If system pressure begins to exceed that specified by the manufacturer, it offsets the spring, and a ball check valve opens and recirculates the fluid to a point outside the pressure-controlled area for reuse.

A pressure relief valve is primarily a backup system. It is set to open at some specific, higher-than-normal, system pressure to act as an escape valve. In normal operation the valve is never used. Should something cause the system to experience a higher than allowable pressure, its force will be greater than the spring tension in the valve, causing the valve to open and allowing the excess pressure to vent.

Fluid Control Valves

Similar to electrical systems, hydraulic systems have their own kind of fuses and diodes in the form of valves, though fuses are more common in large systems that support several functions. If there is a rupture in a line, the fuse prevents fluid loss and allows continued activation of other hydraulic components. Check valves permit flow in one direction only, preventing fluid from backing up in the system. Thermal relief valves are activated by excessively high fluid temperature, which causes fluid to expand. The valve will eliminate the excess fluid.

Actuators

The purpose of an actuator is to transform the energy of fluid flow to a mechanical force. There are several types of basic actuators: single acting, double acting, and rotary. The single-acting cylinder depicted in Figure 14-3 moves under hydraulic pressure in only one direction. To return to its original position, it requires some outside force such as a spring. Some flap systems are single acting, using the air load to retract the flaps.

The double-acting cylinder (nondifferential type) uses the same force in both directions, because each side of the cylinder has an identical surface area, as illustrated in Figure 14-4. Generally, the left rod in the illustration will not be connected to anything, but it is there to take up space and assure equal surface (equal pressure) on both sides. Cessna uses the simple rack and rotary-actuator-driven pinion to retract the main landing gear on its single-engine aircraft. For tasks requiring continuous motion, a piston-and-vane type hydraulic motor is used. Such a motor, common on larger aircraft, provides high power output with instantaneous reversal capability.

The alternative to a hydraulic actuator is an electromechanical actuator. For all the talk within the aerospace power and motion control industry about electromechanical systems, it is somewhat surprising that they are still fairly rare in the industry and essentially nonexistent on light, general-aviation aircraft. At least for the time being, these devices appear to be relegated to a secondary or backup system status on a few select military and air transport aircraft.

There are two primary methods of incorporating electromechanical actuators. One method is a system controlled by a single, main motor with flexishafts that go to the in-

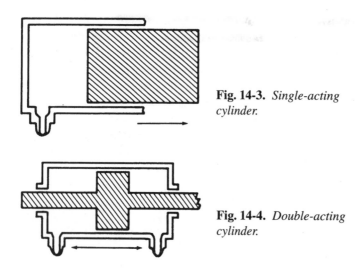

Fig. 14-3. *Single-acting cylinder.*

Fig. 14-4. *Double-acting cylinder.*

dividual devices they will power. This is analogous to an octopus with its numerous arms reaching out all over the aircraft performing various tasks.

The second method is a motorized actuator dedicated to each specific device, thereby eliminating long-running flexible shafts coming from a single motor. For instance, there might be a single, dedicated motor with direct drive shaft on each aileron. The advantage is the reduced likelihood of loss of control of a given device due to a damaged flexible shaft in the event of catastrophic failure elsewhere on the aircraft. And while there appears to be a slight preference in the military and air transport markets toward local motorized actuation, it is unlikely that electromechanical systems will take over hydraulic systems in the near future on any class of aircraft.

SYSTEM APPLICATIONS

Of the many different types of hydraulic systems, the most common is the independent brake system. Older, light aircraft use a single-unit diaphragm master cylinder and brake actuator. The master cylinder contains the fluid, and when the pilot pushes on the pedal, the fluid fills the wheel cylinder and applies pressure to the brake. Hydraulic brake systems on newer, light aircraft, and on all larger aircraft, require a reservoir to hold more fluid and to compensate for temperature change.

Take, for instance, Piper's PA 28-161 Warrior II. Because this aircraft has fixed landing gear, there is a very simple hydraulic system to operate the toe, hand, and parking brake system. It is a highly effective system requiring minimal maintenance. The reservoir, located on the top left, front face of the firewall, is easily accessible for preflight inspection. The system works so well, it is essentially duplicated in other larger Piper aircraft, such as the twin-engine PA 34-220T Seneca III, which has a separate hydraulic system for its retractable landing gear.

On the opposite end of the spectrum is the pressurized system shown in Figure 14-5. System pressure is maintained by an accumulator, essentially a metal sphere. The sphere is split in half by a rubberlike diaphragm. On one side is a dry-air precharge; on the other is system fluid. In addition to absorbing system shocks, the compressible air allows pressurization of the system as fluid pushes against the rubber diaphragm. An automatic unloading valve, which senses system pressure, locks the system, trapping the pressure, and reroutes the continuous stream of fluid from the pump back to the reservoir. As system pressure decreases, the valve senses the reduction, opens, and allows the pump to continue the flow of fluid into the system until the pressure builds up again. While effective, such a system is very costly, complex, and has a greater tendency to wear and leak.

The open center system in Figure 14-6 is found most commonly on light, general-aviation aircraft. It has no accumulator or constant system pressure. When there is no demand on the system, fluid travels from the pump through the open center of each selector valve and back to the reservoir. When hydraulic power is required, the valve rechannels the fluid to the actuator and the fluid from the opposite side of the actuator goes to the reservoir. This is a much less costly and simpler system and is better suited to light aircraft.

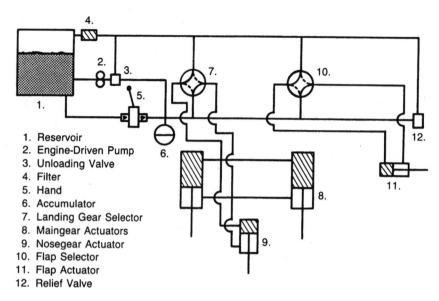

1. Reservoir
2. Engine-Driven Pump
3. Unloading Valve
4. Filter
5. Hand
6. Accumulator
7. Landing Gear Selector
8. Maingear Actuators
9. Nosegear Actuator
10. Flap Selector
11. Flap Actuator
12. Relief Valve

Fig. 14-5. *Pressurized system.*

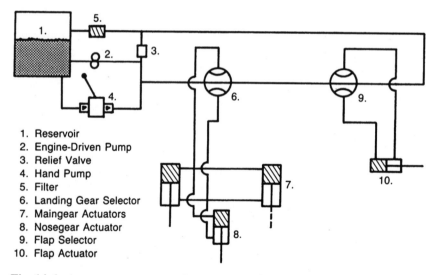

1. Reservoir
2. Engine-Driven Pump
3. Relief Valve
4. Hand Pump
5. Filter
6. Landing Gear Selector
7. Maingear Actuators
8. Nosegear Actuator
9. Flap Selector
10. Flap Actuator

Fig. 14-6. *An open-center system has no accumulator or constant system pressure.*

PREFLIGHT INSPECTION

As with all preflight inspections, follow the POH. There are several hydraulic system checks that should be made on all aircraft whenever practical. Check for traces of hydraulic fluid on the ground, the underside of the airplane, and inside the cowling. Where accessible, check the general security of all hydraulic lines, fittings, and actuator cylinders. Always check the fluid for proper quantity and type, especially if it has been changed recently. The wrong fluid will lead to disaster.

It is also a good idea, when in flight, to press the brake pedals as you near your destination airport. If you find they offer no resistance and go all the way back, you might want to check out the availability of the long runway.

PREVENTIVE MAINTENANCE

Hydraulic fluid not only allows the system to do its work, but it also acts as a lubricant to protect it. I cannot overstress the importance that should be attached to the quality of hydraulic fluid. Seventy-five percent of all hydraulic problems are traceable to fluid problems, and most often contamination! Keep hydraulic lines properly secured; they should not be loose in the clamps. Be especially watchful for chafing of the lines where they pass through the bulkheads or near other components. Where accessible, fittings should be hand checked for tightness during preflight.

Hydraulic fluid has four primary enemies: water, air, heat, and dirt. There are two approaches to dealing with these problems. Airframe manufacturers try to engineer the system to prevent or at least minimize their occurrence. The second method is for fluid manufacturers to try to make the fluids themselves resistant to them. Some of the best preventive maintenance you can perform is to try to minimize as much as possible these major hydraulic fluid enemies.

Water

In larger aircraft that use phosphate esters, water mixed in will produce acid, which causes problems for the system. But even in aircraft that use mineral-based hydraulic fluids, water can lead to corrosion of system parts. Care must be taken to keep water out of the system, and be especially watchful with aircraft that operate in humid climates. The manufacturers, for their part, mix additives into the fluid that help to counteract acids and make the fluid resistant to problems.

Air

Air in the system will result in oxidation of the fluid. Fluid is nonfoaming to reduce the potential for air to get into the system, but it is important to check the system periodically to assure that there is no buildup of air. This is commonly done during routine scheduled maintenance such as during a 100-hour inspection.

Heat

The chance of fluid oxidation as the result of air in the system increases as the working temperature of the fluid increases. In general, heat degrades hydraulic fluid. Under normal conditions the system is able to handle heat, but constantly working a high-load hydraulic system such as the gear can cause long-term problems.

Dirt

Hydraulic fluid must remain uncontaminated because it acts as a lubricant for the system. Never, never, never change fluid during dusty conditions, reuse fluid, or store extra fluid in an open or dirty container. Dirty hydraulic fluid is like using sandpaper for a lubricant. At best, dirt in the system reduces component effectiveness and will lead to component and system failure.

TROUBLESHOOTING

There are a few general principles to keep in mind when troubleshooting any hydraulic system. Because pumps make fluid flow, an inexplicable reduction in flow means a pump or pump-drive problem. No flow and no pressure mean no pump, but the absence of pressure alone doesn't necessarily indicate an inoperative pump. Remember that resistance is required to generate pressure. Fluid flow with little or no pressure means there is no resistance. Find out where the fluid is going; it's leaking out somewhere!

If you have system flow and an actuator doesn't move, only one of three things can be happening. The first possibility is that the fluid may be bypassing the actuator through an internal or external leak. Look for traces of a fluid leak. The second possibility is that the fluid may be returning to the reservoir. If this is the case, the culprit will probably be bad seals or a relief valve that is stuck open. If neither of these appear to be the problem, then the only logical reason remaining is a mechanical failure somewhere in the system.

One of the more common hydraulic system problems is a noisy or chattering pump. If you experience this, there are two likely culprits: pump cavitation or the pump drawing air. *Pump cavitation* is the sudden formation and collapse of low-pressure bubbles resulting from pump rotation. There are several possible causes of cavitation: too low an operating temperature, a dirty inlet strainer, an obstruction in the inlet tubing, or too high a viscosity fluid. All of these will cause the fluid to move more slowly than it should, causing the pump to cavitate. A pump that's drawing air does so if there is insufficient fluid, a leak in the intake tube, a bad pump shaft seal, or if the fluid is foaming in the reservoir. There are also any number of actual pump parts that can fail and cause similar symptoms, but low fluid level is the most common cause.

If the system is overheating, first check to see if the heat exchanger is clogged. Otherwise, a relief valve may be operating continuously, creating excessive heat, or the wrong viscosity fluid could have been put into the system. Slow or erratic pump operation indicates air in the fluid, internal leakage in an actuator, or simply a bad pump. Low system pressure could be a result of dirty fluid, a relief valve stuck open, or the pressure control set too low. Absence of pressure could be the result of a relief valve that is stuck open, a faulty actuator bypassing fluid or, most probably, insufficient fluid in the system. None of these, with the exception of a low fluid level, are easily diagnosed by the pilot, and any such problems should be referred to a mechanic.

Spongy actuation, most noticeable in brakes, is the direct result of air in the lines. Bleeding the lines of the trapped air will cure the problem. Finally, if the hydraulic pressure-gauge needle is bouncing, there is air in the gauge line. By slightly loosening the connection where the line meets the gauge, the air will escape. If all this has sounded devastatingly confusing and complicated, you may take some consolation from the fact that if you keep the fluid clean and the level up, you can avoid 75% of all hydraulic problems.

15
Pneumatic Systems

AIRCRAFT PNEUMATIC SYSTEMS USE AIRFLOW TO PRODUCE EITHER vacuum or pressure for driving gyro instruments, operating de-icing boots, maintaining cabin pressurization, and performing other pressure-related chores.

Aircraft of a few years ago were equipped with venturis mounted on the fuselage in line with the propwash, as shown in Figure 2-14. A venturi is basically a tube flared at the ends and constricted in the middle. As the speed of the air passing through the constricted part of the tube increases, the pressure of the air decreases, thus creating a vacuum, as illustrated in Figure 15-1. The venturi-driven vacuum system is relatively effective for cruise flight but highly susceptible to icing and virtually unusable during ground operations.

WET-PUMP SYSTEMS

To solve the icing problem, an engine-driven vacuum pump system was developed. The pump, shown in Figure 15-2, is impervious to icing, is mounted on the accessory section of the engine, and uses engine oil for both lubrication and cooling; hence they are called "wet" pumps. Air from the cabin is pulled through the gyros and into the pump, which is lubricated with engine oil. An oil separator then returns most of the oil to the engine and exhausts the air and any residual oil out of the system.

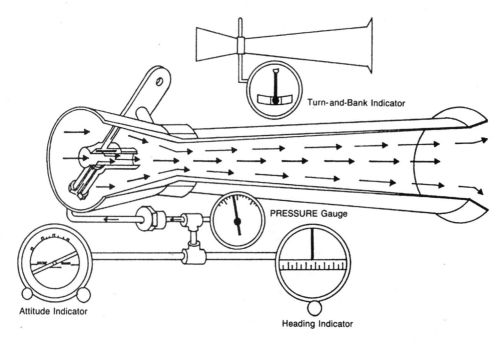

Fig. 15-1. *Venturi vacuum system.*

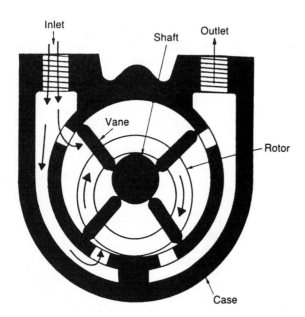

Fig. 15-2. *Vane-type engine-driven pump.*

The system works but has disadvantages. It is relatively heavy because it requires an oil separator. It is also complicated, and oil sometimes finds its way to the wrong part of the system, causing contamination of the de-ice boots and control valves. And the oily exhaust air always leaves a streak on the underside of the fuselage.

Probably the biggest drawback of the wet-pump pneumatic system is that instrument flying is limited to lower altitudes. The problem is that the low-density air found at high altitudes is insufficient for the pump to create enough vacuum to drive the gyros. If the wet-pump is used in high-altitude operations, the gyros and the rubber de-icing boots, located on the pressure or downstream side of the pump, could be ruined by oil contamination.

DRY-PUMP SYSTEM

The solution to the many problems of the wet-pump came with the invention of—you guessed it—the dry-pump system. The simple, lightweight, dependable, and self-lubricating dry-pump system has no oil contamination or cooling problems. Because it can power either a vacuum or pressure system, it is able to drive the gyros and the de-ice boots as well as pressurize the aircraft door seals.

VACUUM VERSUS PRESSURE SYSTEMS

The most common type of aircraft pneumatic system is the vacuum system shown in Figure 15-3. Air from the cabin is drawn into the system through a central air filter. It then goes through the gyro instruments, past the vacuum (suction) gauge, and through the relief valves. At that point it goes into the vacuum side of the pump, out the pressure side, through an air/oil separator, and is discharged overboard.

Twin-engine aircraft vacuum systems work essentially the same with one pump per engine. This provides a safety margin because either pump alone creates sufficient vacuum for normal operation. Both pumps connect to a common manifold and share the same tubing and valves.

Vacuum systems are more common than pressure systems in light aircraft because these aircraft generally operate at lower altitudes. The vacuum side of the pneumatic

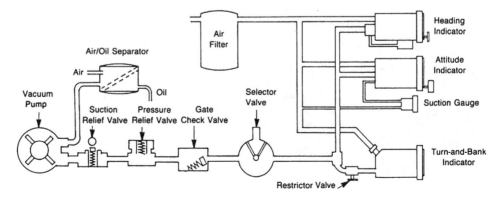

Fig. 15-3. *Typical pump-driven vacuum system.*

Chapter Fifteen

pump becomes less and less efficient as ambient air density decreases. Using a vacuum system in aircraft that routinely operate at higher altitudes results in shortened pump life because the pump has to work harder in the thin air. Smaller aircraft require less vacuum action from the pump because it usually only powers the gyros. A typical twin-engine vacuum system is shown in Figure 15-4.

Pressure systems can move more air at higher altitudes than their vacuum counterparts. Air enters the pressure system through a cabin inlet filter, goes directly into the intake (vacuum) port of the pump, and out the pressure port to the pressure-regulating valve. After going through another filter, the air is routed to the gyro instruments, past the gauge, and finally is released overboard.

Twin-engine aircraft have one pressure pump per engine and, as in the vacuum systems, they share common tubing and valves. The problem is that pressure systems tend to create moisture, especially in high-humidity locations such as operating near oceans, lakes, rivers, and in areas with heavy rain. Because moisture in the lines and gyros can cause significant problems, filters of aircraft operated in high-humidity areas should be changed more frequently than recommended by the manufacturer.

The dry-air pump is the heart of the modern pneumatic system. The pump rotor has self-lubricating carbon vanes that are specially designed to wear and lubricate the pump

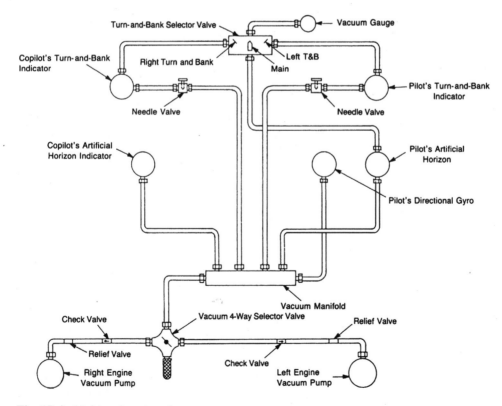

Fig. 15-4. *Multiengine aircraft vacuum system.*

at a set rate. Unlike older, bidirectional pumps, the dry-pump turns in only one direction. To provide maximum life, the vanes are canted at a carefully calibrated angle, making it essential that the pump be installed correctly. A pump that has been dropped should never be put into service. The carbon vanes and rotor are surprisingly fragile, and while there may be no visible sign of damage, internal damage may have occurred.

For gyros to work properly, exact vacuum (or pressure) must be maintained, despite the potential for system fluctuations caused by power surges, de-ice-boot inflation, and engine-speed changes. The gyro instruments are protected from excessive vacuum by a regulating valve. Under normal operating conditions, the pump draws more air than is necessary to operate the gyros, so the valve regulates the airflow by drawing from an alternate air source that bypasses the gyros. Similar to the vacuum system, the pressure system has a regulation valve that serves to relieve pressure by allowing excess air pressure to escape prior to entering the gyro instruments.

The vacuum gauge needle responds directly to the amount of suction in the tubing that passes by the gauge. In twin-engine aircraft, where there is one pump per engine, the gauge has a pump-failure alerting mechanism. Inside the vacuum gauge—out of the sight of the pilot—two small red balls are held independently by the suction of each engine-driven pump. If, for instance, the left engine fails, its corresponding pump will also fail and there will no longer be suction on the left red ball. A small spring, normally held back by the suction, now pushes the ball into the view of the pilot. Pilots must be cautious, however, as this is not necessarily a reliable indication of engine failure. The windmilling propeller of a failed engine may keep the vacuum pump operating normally. In that case, the red indicator would not appear until the pilot feathers the prop.

A pressure gauge works essentially the same as a vacuum gauge in both single-engine and multiengine aircraft. Both types of gauges should read approximately in the middle of the green arc during normal operations.

Modern fittings and tubing are designed for maximum airflow. While the tubing in older pneumatic systems often had sharp, 90-degree turns, modern systems are engineered with less acute bends. Even the design of the tubing itself is geared for minimum flow resistance because hose cracks, twists, crimps, and improper internal dimension can lead to significant airflow problems and may seriously decrease the life of the pump.

Pneumatic system filters capture impurities in the air, safeguarding the gyro instruments. Clean air is essential for unrestricted airflow, protection of delicate gyro instruments, and pump longevity. Most gyro problems can be traced back to system contamination. At the head of the list are dust, women's face powder (avoid the need to powder your nose before landing), cigarette smoke, dirt, some cabin-cleaning agents, and dry fire-extinguishing agents. It is the job of various filters to protect the system from these contaminants.

The inlet air filter is the primary system filter; it removes most major pollutants from the air before they enter the system. The vacuum system has a garter-type foam inlet filter that cleanses the air as it enters the regulating valve. In the pressure system, the inlet filter is located just upstream from the pump itself. An "in-line" filter, which comes between the pump and the instruments, is used in the pressure system to trap the carbon lubricating particles before they enter the gyro instruments.

Chapter Fifteen

PREFLIGHT

Preflight of the pneumatic system is limited to an under-the-cowling check of the general integrity of as much of the system as is visible. Hoses and fittings should appear in good condition with no kinks or twists. If the pump is visible during preflight, check the base where it attaches to the engine; traces of oil on the accessory drive pad indicate a bad seal. During the cockpit preflight, be especially watchful if the vacuum pump has passed the 500-hour mark, or 200 hours for the pressure pump.

The vacuum/pressure gauge should be an integral part of your instrument scan. Be aware of the normal gauge reading for cruise flight, as deviations may indicate impending failure. Any excessive gyro precession during ground operations is reason to suspect system problems. For instance, the attitude indicator should not show more than a 5-degree bank during taxi turns, assuming, of course, that you're not taxiing so fast that you actually do bank more than 5 degrees when you turn.

PREVENTIVE MAINTENANCE

The single most effective preventive maintenance a pilot can perform on the pneumatic system is to forbid smoking in the aircraft. Tobacco smoke is a prime cause of clogged filters and shortened gyro life. One of the worst and most disgusting tasks on a 100-hour inspection is to clean the filter of an aircraft that is regularly flown by smokers. It is literally coated with a thick, mucouslike layer of nicotine. Collect some and put it in a jar, and the next time someone wants to light up in your airplane, scoop out a big glob, hold it under his or her nose, and shout, "That's what's in your lungs!" It's a good idea to have a sic-sac ready in advance.

Where gyros are concerned, small problems rapidly turn into big ones. Some mechanics, for instance, check filters at the required time and if they don't look dirty, they won't clean them. The problem is that dirt gets trapped inside the filter where it's not visible, then it works its way out and into the gyros.

Literally any small particle can cause a gyro to fail! In one general-aviation airworthiness alert, the gyros stopped working when the vacuum system had excessive suction. The master air filter appeared clean, but after it was torn open, it was discovered that water had leaked inside. The filter had absorbed the water, swelled, and blocked the airflow. From the outside, the filter looked fine.

In general, filters should not be cleaned; they should be replaced. The problem is more subtle than just potential contamination. The longer a filter screens, the lower the volume of air that can pass through it. If the filter is located upstream from the pump, the decreased airflow forces the pump to work harder. This increase in workload causes the pump to run hotter, which increases the wear on it and shortens its life.

Under normal conditions, the vacuum regulator filter or pressure system inlet filter should be changed every 100 hours of operation or every year, whichever comes first. Central air filters and in-line filters should be replaced every 500 hours or annually, whichever comes first. If an aircraft is operated in a high-humidity area, mechanics often recommend changing in-line filters every 300 hours.

If you install a new pump, always replace the old gasket with the new one that is provided. Used seals have been compressed and are not designed to be reused. The gasket seals the pump to the accessory-drive-pad oil hole and also serves as a heat shield. A leaking seal will cause the system and pump to be contaminated with oil. A new pump also warrants a check of the system and change of all filters. This should include cleaning the pressure lines from the pump to the instruments because carbon particles may find their way through the filter and begin to collect there. Also clean the inlet lines.

If there is any drawback to the dry-pump system, it is the potential for carbon contamination. While this problem may be obvious in the pressure system, some mechanics think the vacuum system, with its pump downstream from the gyros, also may be vulnerable to carbon contamination. They hypothesize that as the pump shuts down, the instrument cases momentarily have a lower pressure than the pump, causing a reverse airflow that could pull carbon dust back into the instrument case. Whether or not this actually happens, the best safeguard is to change filters regularly and keep the system clean.

The lining of deteriorated hoses may break away and flow through the system until it clogs a filter, lodges in a valve, or jams the pump. The effect of a collapsed, kinked, or twisted hose is similar to that of a clogged filter. The restricted airflow raises pump operating temperatures, which results in a shorter pump life. Damaged or loose fittings, in addition to acting as another air source that may bypass gyros, also can allow engine cleaning solvents to enter the pneumatic system. When this happens, the solvent can mix with carbon particles to form sludge. In the pump, sludge will cause the coupling to shear! As a precaution, never direct high-pressure solvents on a pneumatic system component when cleaning an engine. In fact, it is a good operating practice to encase these components in plastic bags before using solvent cleaners. A word of caution, however: The bigger the bag, the less likely you will be to miss it on a preflight, if you forget to remove it.

Another problem associated with loose fittings or leaking hoses is the reduced pressure, which results in decreased gyro airflow. The pilot, noticing a change in gauge reading, may have a mechanic adjust the regulator. This makes the pump work harder and run hotter to compensate for the air loss, leading to shorter pump life. It is cheaper in the long run to replace bad fittings and hoses than to buy a new pump.

TROUBLESHOOTING

Any of the following trouble signs warrant checking the entire pneumatic system:

1. The pump fails soon after installation.

2. The aircraft has a history of short pump life.

3. The vacuum (or pressure) indication is above or below proper level.

4. Gyro performance is erratic.

5. The de-ice system malfunctions.

6. A door-seal system malfunctions.

7. An autopilot system malfunctions.

Chapter Fifteen

Early pump failure is most often caused by the wrong pump being installed; check the part number of the pump. Otherwise, a thorough investigation of all pneumatic components is in order, including gyros, de-ice boots and valves, door seal valves, and the pneumatic autopilot system. Contamination can overload a pump, causing it to fail early, sometimes within hours of installation! If there are high, low, or erratic gauge indications, check for hose problems, clogged filters, oil in the system, or loose fittings. If absolutely everything else checks out, then it could be a simple regulator-setting problem.

Excessive gyro precession is probably caused by dirty filters. If the problem continues after changing them, you almost certainly have a bad gyro. If the gauge indication varies with engine RPM, it is the result of regulator-seat contamination, which prevents constant pressure as the pump speed varies with engine RPM. If you find the mechanic is frequently making small regulator adjustments to correct the gauge reading, the filters probably are becoming clogged. This is particularly common in aircraft that carry smokers and eventually will result in premature pump failure. But replacing a pump doesn't necessarily mean the system problem has been cured.

Short pump life should be like a red flag to a pilot or mechanic, as it indicates that there is something wrong with the system. I knew of an aircraft owner who experienced seven pump failures, one after only six hours of operation! Finally, the pilot contacted the manufacturer directly and the problem, which was related to the de-ice boots, was cured in one day.

Once in flight, there is nothing the pilot can do about a pneumatic-system failure. The simple fact is pump rotors are designed to break if there is any trouble; otherwise they might cause an engine problem. While twin-engine aircraft pilots have the edge afforded by redundancy, the single-engine aircraft pilot is typically backed against a wall in the event of system failure. For the single-engine pilot who must rely on an aircraft for business, it may be worthwhile to have a standby vacuum system (SVS) installed. One such SVS uses engine manifold pressure, which is diverted to the instruments by a valve connected to the cockpit with a push/pull cable. It operates on the differential between the engine manifold pressure and ambient atmospheric pressure. An SVS offers many advantages, as it is simple and requires no electricity or maintenance.

Glossary

alternating current (AC) An electric current that periodically changes direction of flow and constantly changes magnitude.

ammeter Instrument used to measure current flow.

ampere (amp) Basic unit of current flow (flow rate); an indicator of the passage of electrons through a conductor. One amp is the amount of current that flows when a force of 1 volt is applied to a circuit with a resistance of 1 ohm.

ampere-hour (amp/hour) A rating given to a battery indicating potential duration of the current flow under ideal conditions. It is the quantity of electricity that passes through a circuit if 1 amp has flowed for 1 hour (amps × hours).

brake horsepower Horsepower produced by the engine minus losses due to friction, exhaust, and cooling.

bus bar (bus) Power distribution point, usually a metal strip where several circuits are connected.

circuit A number of conductors connected together to complete an electrical path.

circuit protection Devices in a circuit that protect wiring and/or appliances, such as fuses and circuit breakers.

current The movement of electricity through a conductor.

detonation Explosive, near-instantaneous release of fuel heat energy due to fuel/air mixture reaching its critical temperature and pressure.

direct current (DC) Electric current that always flows in only one direction.

direct drive Propeller bolted to, and turned at the same speed as, the crankshaft without reduction gearing.

GLOSSARY

factory remanufactured engine Engine is completely disassembled and rebuilt by original manufacturer to factory-new tolerances and zero time.

fuse A metal link that melts when overheated by excess current; used to break an electric circuit whenever the load exceeds a predetermined maximum.

inverter An accessory that converts direct current to alternating current.

major overhaul Complete engine disassembly, inspection, and overhaul to manufacturer's specification. Total engine time in the logbook continues.

master switch A pilot actuated switch designed to control all electric power in the aircraft.

normally aspirated Engine-rated power lacks ability to maintain sea level rated power at altitude.

open circuit A break somewhere in a conductor preventing current flow.

parallel circuit Two or more circuits connected to the same power source and ground.

preignition Premature ignition of fuel/air mixture in cylinder prior to spark plug discharge.

rated power Maximum continuous horsepower output when operated at specified RPM and manifold pressure.

rectifier An accessory that converts alternating current to direct current.

relay An electric switch that is operated by either an electromagnet or a solenoid.

series circuit A circuit in which the current must flow through all the circuit elements in order, by a single path.

short circuit Contact between conductors permitting a short, low resistance path back to the power source.

time between overhaul (TBO) Manufacturer recommended period to first major overhaul; assumes compliance with recommended operating procedures, maintenance, and inspections.

top overhaul Entails removal of cylinders, deglazing cylinder walls, installation of new piston rings, and touch-up of valves and valve seats.

volt A unit of electromotive force (voltage); a measure of electrical pressure.

voltage regulator An accessory that maintains a constant level voltage supply despite changes in input voltage or load.

watt Unit of electric power, a rate of doing work. In a direct current circuit: watts = voltage × amps.

Index

INDEX

INDEX

INDEX

INDEX

turbocharger systems (*Cont.*):
  Cessna TU206G turbocharger, **145**
  changing the oil, 150
  contaminants in cooling fluid, 148, 150
  control of turbocharger, 145
  cylinder head temperature (CHT)
      gauges, 147–148
  cylinders, 144–145
  exhaust gas temperature (EGT)
      gauges, 147–148
  failure, emergency procedures, 150
  flow diagram, **144**
  fuel-air mixture, 143, 144–145, 147, 148
  heat problems, 147–148
  high-altitude operation, 144
  historical development of super- and
      turbo-charging, 142
  impeller, 143
  leaks, 150
  lubrication, 148, 150
  maintaining sea level power, 142–143
  manifold pressure (m.p) gauges, 146
  misconceptions, 144–145
  overboosting, 146, 150
  pistons, 144–145
  rich-lean setting, 143, 144–145, 147, 148
  superchargers, 101, 141–142, **142**
  time between overhaul (TBO), 145
  troubleshooting, 148–150
  turbine wheel, 143
  vibrations, 150
  wastegate controller, 145–147
  wear on engine, 144–145
turn indicators, 42–44, **43**
  errors, acceptable ranges, 51
  preflight inspections, 49
turning error, magnetic compass, 57
type certificate data sheets (TCDSs), 68

ultra high frequency (UHF) antennas, 25
ultraviolet light damage to acrylic
    transparencies (windows), 17–18

vacuum-pump driven gyro systems, 45, **46, 47,** 50, 61, 283–284, **283, 284**
valves, 111
vapor lock, 87, 108, 131, 139
variable isobaric systems, pressurization systems, 249–251
variation, magnetic, lines of, 58–59, **58**
velocity (air) cooling, 82–83
vents, fuel tank, 134–135
venturi-driven gyro systems, 44–45, **44,** 47, 49, 50, 281
vertical velocity indicator (VVI), 35, **35, 36,** 39
very high frequency (VHF) antennas, 25
vibration damage, 5–6, 112
  friction error in altimeters, 33
  propellers, 214
viscosity of oil, 110, 111, 115–116, 121–122, 124
voltage regulators, 174
voltmeters, 174
VOR/LOC, 25

warranty work, engines, 65–66
wastegate controller, turbocharger systems, 145–147
water damage to acrylic, 18
water in fuel, 134, 135–136
water in hydraulic fluid, 277
weather, airframe inspection, 4
Weeping Wing anti-icing system, 266–268, **267**
wet sump oil/lubricating system, 118–121, **119, 120**
wet-pump pneumatic systems, 281–283
wheel wells, corrosion, 7
windows (*see* transparencies)
wobble fuel pumps, 136
Wright 760 radial engine, **78**
wrinkles in exterior skin, 5

xenon flash tubes, strobe lights, 13

About the Author

David A. Lombardo is a professional pilot and an international aviation training consultant, specializing in simulation, training program design, and organizational development. He is a former assistant dean of the Division of Aviation at Lewis University and former Aviation Program Director at Bowling Green State University. He has flown 34 different types of aircraft in thousands of hours in the air. His FAA certifications include: Airline Transport Pilot; CFI, airplane single- and multiengine, and instrument; and Airframe and Powerplant Mechanic. He has lectured widely on aviation and aviation education.